Dinosauria And Prehistoric Creatures Magazine

Fall 2022

I0430304

- How Plesiosaurs Swam: New Insights into Their Underwater Flight Using "Ava", a Virtual Pliosaur
- The stunning Art work of Vito Fabrizio Brugnola
- Scansoriopterygids, Anchiornis, and How Flight May Have Actually Started
- Skorpoiventor Real or not?
- Atrociraptor
- The unusual Diplocaulus
- Interview With Evan Johnson-Ransom

DINOSAURIA AND PREHISTORIC CREATURES MAGAZINE

Fall 2022

CONTENTS

DINOSAURS * PREHISTORIC CREATURES * FOSSILS

75 THE STRANGE AND UNUSUAL DIPLOCAULUS BY SHETAN NOIR

03 THE STUNNING ARTWORK OF VITO FABRIZIO BRUGNOLA

By shetan noir

12 SCANSORIOPTERYGIDS, ANCHOIRNIS AND HOW FLIGHT MAY HAVE ACTUALLY STARTED.

By Christian DeRusso

31 SKORPIOVENATOR, REAL OR FAKE

By Dr. BRIAN CURTICE

39 HOW PLESIOSAURS SWIM

*By Max Hawthorne,
Mark A. S. McMenamin
Paul De La Salles*

Interview with Evan Johnson Ransom Author of Dinosaur World **93**

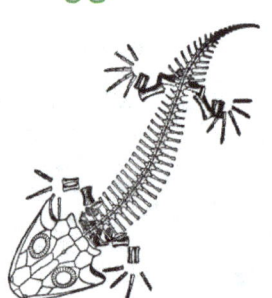

70 ATROCIRAPTOR

By Dr. Brian Curtice

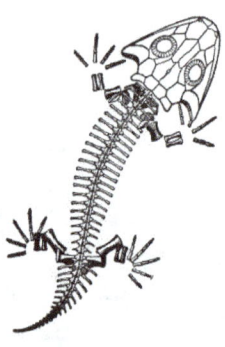

90 JURASSIC JOURNEY

By Shetan Noir

 DINOSAURIAMAGAZINE@YAHOO.COM

HTTPS://SHETANNOIR.WIXSITE.COM/SQUATCHGQMAGAZINE

THE STUNNING ARTWORK OF VITO FABRIZIO BRUGNOLA

patreon.com/vfb_paleoart

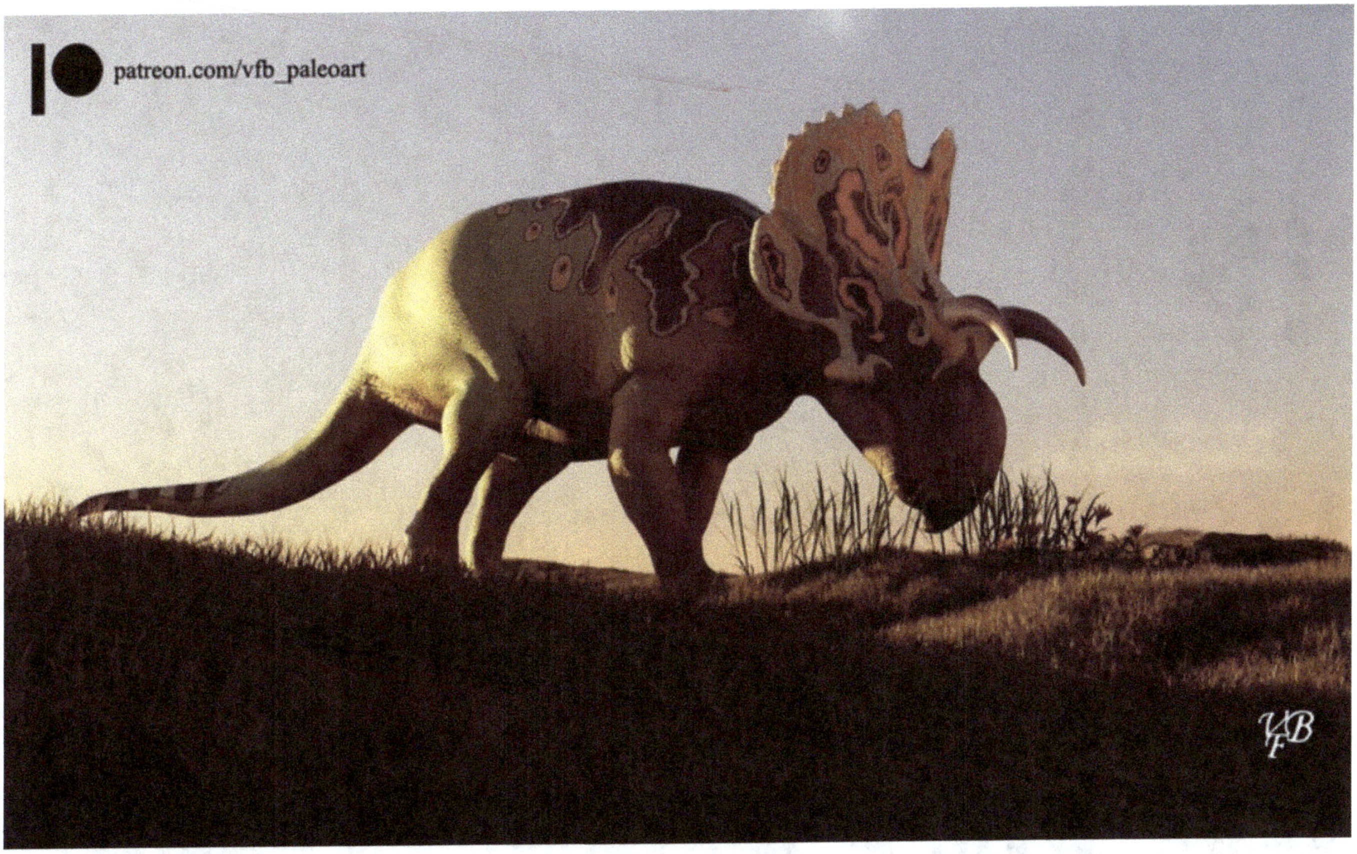
patreon.com/vfb_paleoart
VFB

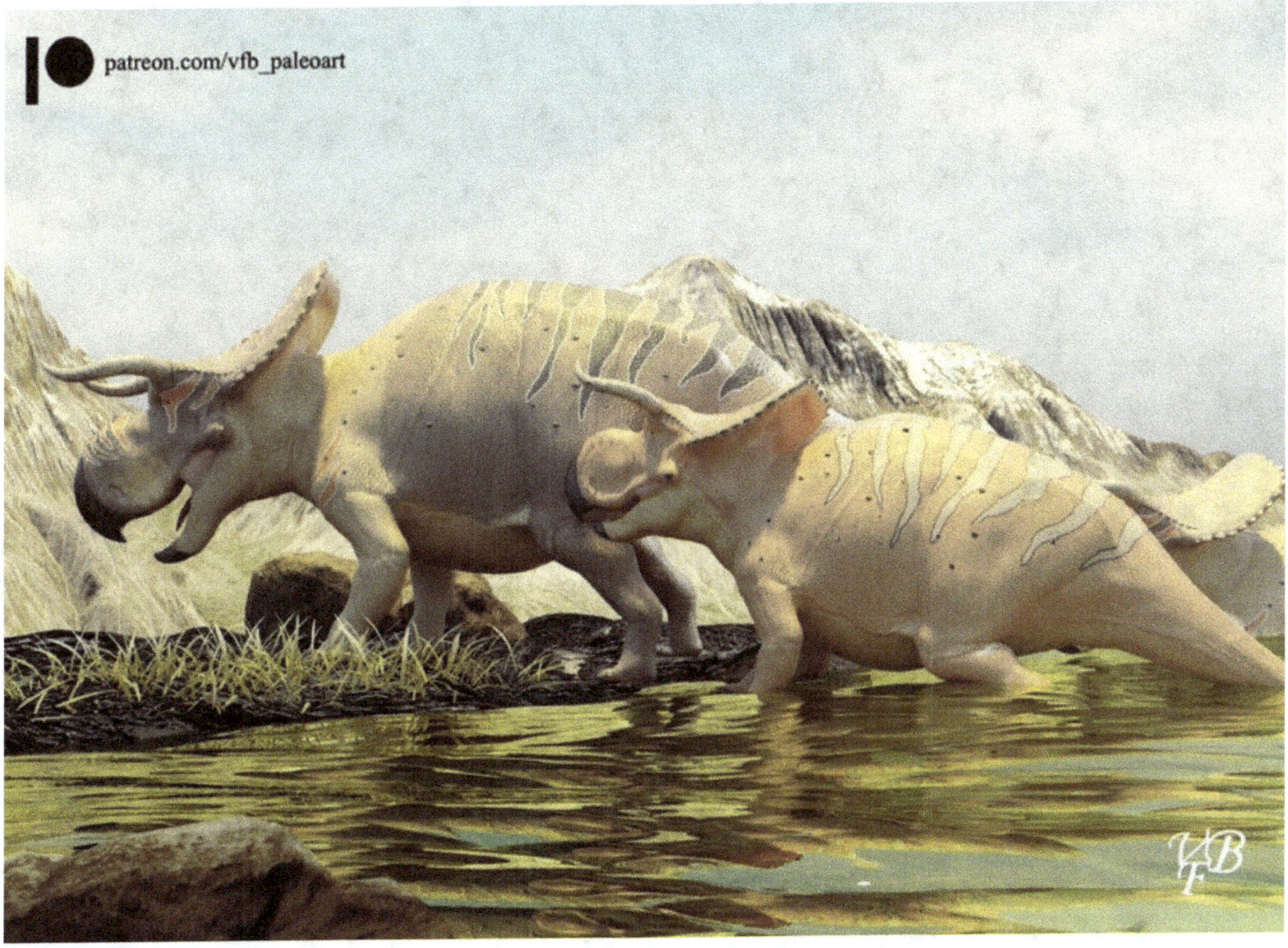
patreon.com/vfb_paleoart
VFB

https://www.palaeoafterdark.com/about

Scansoriopterygids, *Anchiornis*, and How Flight May Have Actually Started

By Christian DeRusso

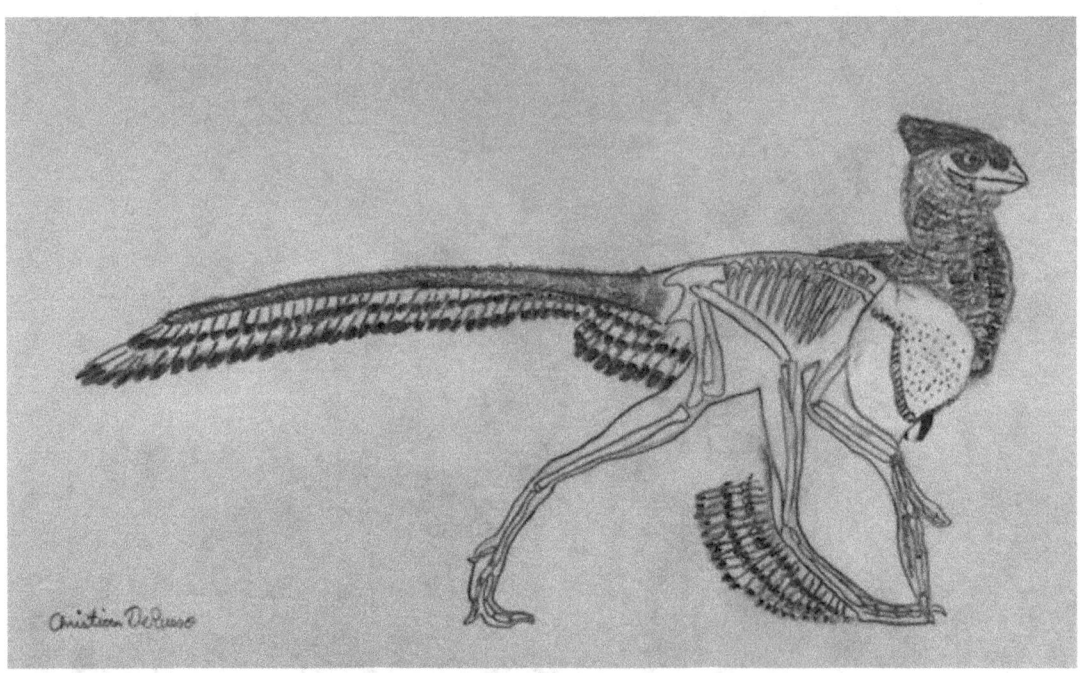

Anchironis huxleyi
(Feathers on the tail and hindlimb shown; head fully skinned with the crest of Pileated woodpecker (*Dryocopus pileatus*))

Introduction

We all know for sure that the dinosaurs' closest living descendants are the birds, and we know birds are the only group of dinosaurs to achieve full powered flight. But as we look more and more into the fossil record and discover new extinct species, our understanding improves. However, new discoveries and new studies in recent years also give us a lot more questions than answers. Even when we find theropods preserved with feathers and, in rare cases, soft tissues. I find it completely condescending to think we thought we could get an answer from looking at the bones and from discovering new species. And yet, the overall evolution of flight in birds is still far from complete. More discoveries and new research have put what we thought we know about the origin of avian flight to the test, and with new ideas changing, I will tell you an even bigger picture about flight being more like an evolutionary trail-and-error created by the forces of nature. An experiment throughout the second of the Mesozoic era as larger and complex that drives the ancestors of modern birds to great length. This section discusses some of the early paravian theropod dinosaurs and oddballs, called scansoriopterygids. A bigger picture of these animals will give us not only a better understanding of the different approaches to how flight evolved. It will also give us a sense that those different approaches tell us flight evolved more than once in dinosaurs.

Scansoripoterygidae

Morphological Comparisons and Constraints

Compared to *Archaeopteryx,* the total length of the arm down to the distal end of the middle digit is slightly shorter in scansoriopterygids. Despite the arm length nearly equal to that in *Archaeopteryx*, the arm has different proportions, (especially in regard to its smaller deltopectoral crest and robust, slightly shorter radius and ulna); these shorter forelimb bones represent a basal plesiomorphic condition. The relative lengths of the proximal phalanges to the penultimate phalange (located in the long outer digit IV) are another non-theropod feature according to Feduccia and Czerkas (2014); the phalanges became progressively shorter distally. This reduction is also believed to be plesiomorphic within Archosauria and is not characteristic of Theropoda. However, a recent paper made in 2020 placed scansoriopterygids closely related to oviraptorosaurs due to morphological similarities (Sorkin, 2020; Motta et al., 2020). The uncertainty with the phylogenetic position of Scansoriopterygidae suggests scansoriopterygid morphology was placed closely to the basal morphology of Pennaraptora section of theropods (Sorkin, 2020). All scansoriopterygids were capable of scansorial locomotion and gliding, but not powered flight, suggesting that the common ancestor of Scansoriopterygydae was also scansorial and volant (Sorkin, 2020). The reason for is this because of the specialized ecology and some morphological constraints these animals had compared to the rest of the theropods within Pennaraptora.

For the scansoriopterygids, the ecology is specialized compared to other theropods throughout the Late Jurassic. For starters, the flight musculature of Scansoriopterygidae was reduced, as it had an underdeveloped deltopectoral crest (DPC), unlike *Microraptor* or *Archaeopteryx* (Huang et al., 2002; Pei et al., 2014), and a small sterna (Wang et al., 2019), unlike the condition seen in *Microraptor* (Xu et al., 2003). Wing loading estimates vary but fall within range seen in extant and extinct gliders ( ), assuming a bat or pterosaur wing model Dececchi et al., 2020). However, none of the models are supported by the anatomical wing loading results; this suggests that the wings of scansoriopterygids may have been a hybrid between several models (Dececchi et al., 2020). Wings of scansoriopterygids, such as *Yi qi* and *Ambopteryx*, are composed primarily of rigid bony elements, thus reducing its flexibility and maneuverability with no specialized muscles or other supporting elements, like elastin within the patagium that control wing shape in bats (Cheney et al., 2017), gliding mammals (Socha et al., 2015), and pterosaurs (Kellner et al., 2010; Palmer, 2018; Yang et al., 2019).

Specialized Ecology

The results made from a study made by Dececchi et al. (2020) suggesting that scansoriopterygids may have been specialists as their gliding capabilities were poor, and the viability to take off put them at a disadvantage compared to gliding animals with more maneuverable and capable flight styles. Their specialized yet poor gliding required relatively high glide speed and average glide ratios that would have been suitable for quickly crossing small gaps in the canopy (Dececchi et al., 2020). This specialized ecology the animal possessed when they were alive likely explains short duration of the fossil record and shows that the gliding ability of scansoriopterygids were a failed experiment in flight origins and avian evolution from theropods (Dececchi et al., 2020). With such highly loaded wings and high speeds of movement, the total cost of climbing to a sufficient launch point to achieve a destined horizontal distance would likely be quite energetically expensive (Byrnes et al., 2011). High loading values reduce the performance values regarding the wings of scansoriopterygids resulting them to be poor gliders as mentioned by McGuire and Dudley (2011).

Scansoriopterygids: The Real "Proavis?"

In his book published in 1926, "The Origin of Birds," Gerhard Heilman made his original description of a species of dinosaur Proavis. He described "Proavis" as an animal with an already developed hind-toe, skull larger in proportion to the skeleton, a specialized hand made up of a second digit longer than the others, forelimbs more spread-out during flight and was said to have an arboreal lifestyle (Heilman, pg. 193-201). However, after reading his book, I had looked upon the animal Proavis and had a hunch that maybe there was something to believe theropod dinosaurs are separate animals and have no relation to the modern birds. Meanwhile, the similar anatomical traits between theropods and birds already indicate (in my outlook) a close relation in terms of relatedness and a common ancestor between them and modern birds. All the recent research on these animals and newer discoveries would have made Gerard Heilman's claim moot. But maybe Heilman might not be off track. As I saw scansoriopterygids

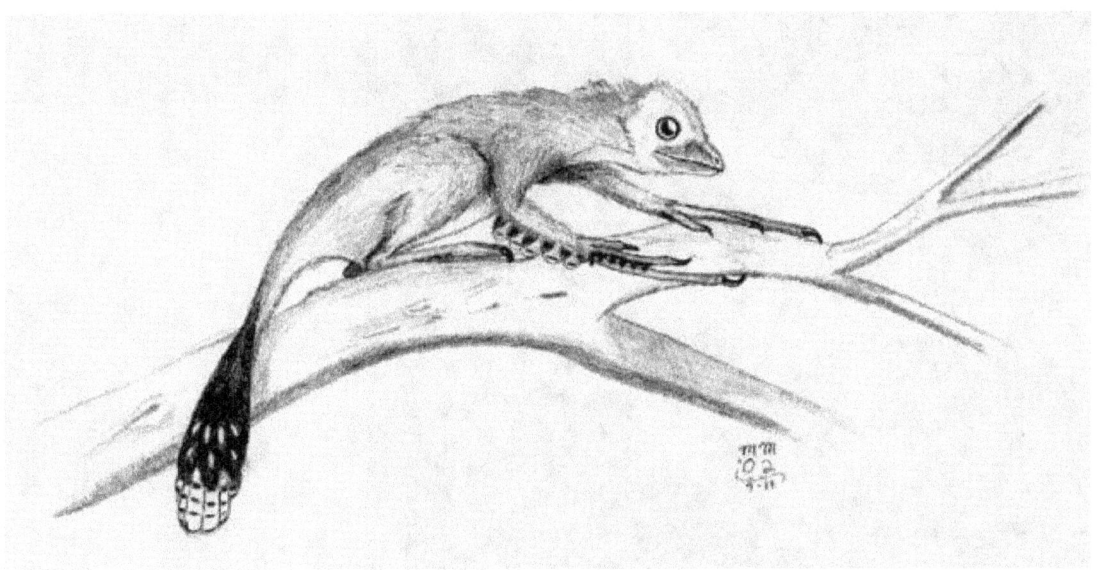

to be evolutionary dead ends when studying flight origins, could "Proavis' be based on an animal yet to be discovered in the future? Could the scansoriopterygids like *Yi qi* and *Ambopteryx* be the real suspect?

Scansoriopterygids had the characteristics of both pretheropods and avian theropods (including the birds) that would seemingly make us believe that they would be the possible reference to the claimed "Proavis" as suggested by Gerhard Heilman. The pretheropod characteristics among the scansoriopterygids include the outer digit of the manus longer than the outer digits, a straight and more robust uter metacarpal that is longer than the mid-metacarpal, and phalanges of the outer manus digit becoming progressively shorter distally (Feduccia, pg. 154, Table 4.2). Meanwhile, there are also avian characteristics present in scansoriopterygids as well and those include long forelimbs that are nearly the length of that in *Archaeopteryx,* the presence of the semilunate carpal, primary feathers on the manus (that is until the discovery of *Yi qi* and *Ambopteryx*), and an anisodactyl foot with a reversed hallux (Feduccia, pg. 154, Table 4.2). Scansoriopterygids would likely fit the description of the "Proavis," although this arboreal lifestyle is now deemed specialized (Dececchi et al., 2020) and they were phylogenetically placed recently near the base of the clade Pennaraptora (Sorkin, 2020). So, I would believe it is safe to say the "Proavis" is still some made up name with no real evidence to back up this.

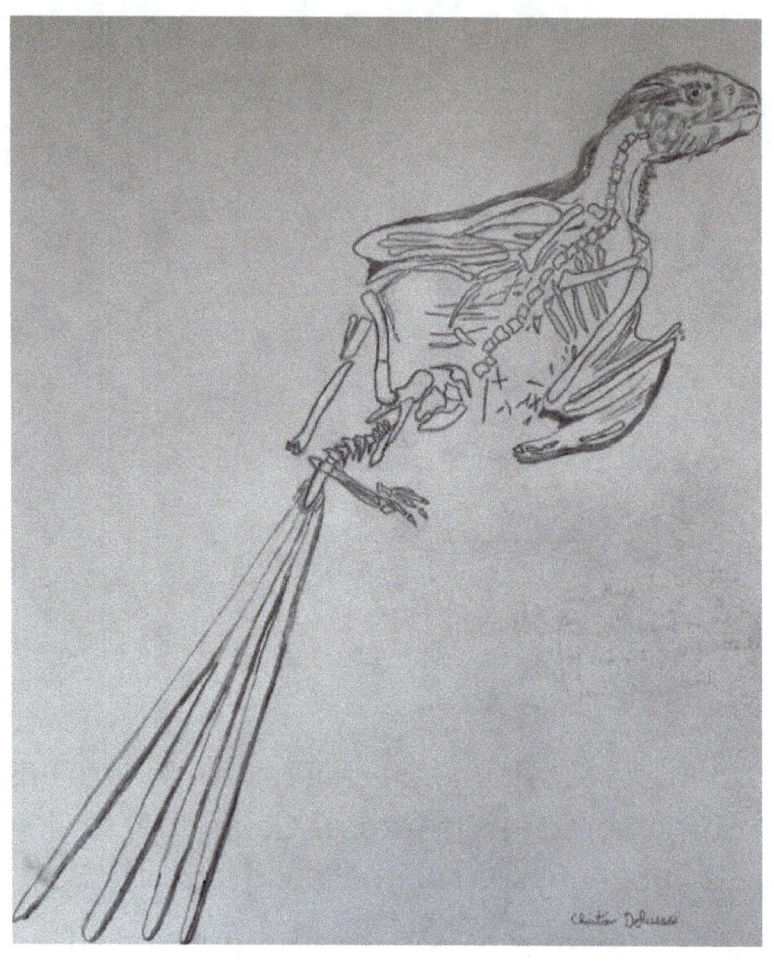

Ambopteryx longibrachia
(Skeleton based on holotype with outlining of the wings and feathers attached to the pygostyle; head fully skinned)

Anchiornithines and Implications of Flight

Anchiornithines are among a recent group of dinosaurs that have added into the debate of how avian flight originated from theropod dinosaurs. They have a set of unique anatomical features have made them separate themselves from other ancestral and derived members of the Pennaraptora clade that may further our understanding of how flight originated. These traits include nutrient foramina on dentary placed in a deep groove, smaller and more numerous anterior teeth that are closely oppressed than there are in the middle, anterior edge of acromion region of the scapula laterally averted or hooked, medial surface of proximal fibula flat, far-shaped posterior dorsal neural spines, extensive large pennaceous feathers on the metatarsals and pes (as also found in *Microraptor* and *Sapeornis*) (Foth and Rauhut, 2017; Agnolin et al., 2019). However, there are traits unique to the anchiornithine group of theropods that worth noting to origin and evolution of flight: the very short deltopectoral crest (DPC) of the humerus and the straight radius and ulna (Foth and Rauhut, 2017; Agnolin et al., 2019). Along with those traits, there are additional skeletal and integumentary adaptations for gliding flight and scansorial locomotion; they are long forelimbs with clawed manual digits, short ilium, distally positioned hallux, and elongated penultimate phalanges of toes III and IV (Hu et al., 2009, 2018; Lefevre et al., 2017, Figures 4d and 5d; Sorkin, 2020, Tables 1 and 2; Sorkin, 2020). These were actually present in taxa such as *Anchiornis* (Hu et al., 2009), *Serikornis* (Lefevre et al., 2017), and *Caihong juji* (Hu et al., 2018). It is from these traits there is a possibility that compared to other members of the Pennaratpora group of theropods, the anatomical traits of anchiornithines may not be well suited for flight and thus would mark the starting trend towards first steps of avian flight. Having a long forelimb consisting of a straight radius and ulna, a short DPC, and a short ilium may mean that their mode of locomotion is likely ancestral even when comparing to

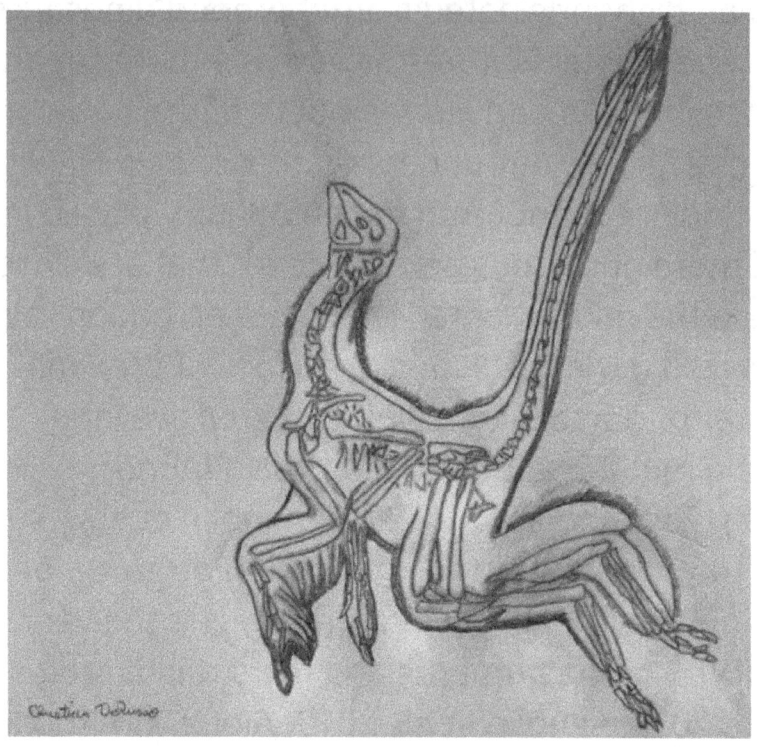

Serikornis sungei in a natural avian death pose.
(Skeleton based on the holotype specimen; right forelimb outlined with symmetrical feathers and propatagium; head skinned exterior outlined with 'dino-fuzz')

Archaeopteryx and other taxa as depicted in recent studies by Wang et al. (2017), Zheng et al. (2014), and Dececchi et al. (2020). Being that the anchiornithines and *Archaeopteryx* both appeared during the Late Jurassic period, the anchiornithines and *Archaeopteryx* may have been the start of a very large trend involving different approaches to how flight originated. These traits only found in the anchiornithines are one of those many approaches that started to occur during the Late Jurassic to Late Cretaceous Period.

Archaeopteryx and other taxa as depicted in recent studies by Wang et al. (2017), Zheng et al. (2014), and Dececchi et al. (2020). Being that the anchiornithines and *Archaeopteryx* both appeared during the Late Jurassic period, the anchiornithines and *Archaeopteryx* may have been the start of a very large trend involving different approaches to how flight originated. These traits only found in the anchiornithines are one of those many approaches that started to occur during the Late Jurassic to Late Cretaceous Period.

The anatomical traits in anchiornithines should be notified as one of the approaches to how avian flight evolved and originated and should be further looked at upon future research. As we look into these group of extinct animals, we must also be very cautious about the flight potential when making a clear picture of how flight originated as well. The reason is that there are certain parts of the anatomy of anchiornithines that would make us want to cautiously examine the potential flight of these animals; they include lack of functionally asymmetrical feathers (char. 1.0), short ulna and humerii (char. 233.0; char. 262.2), and limited pectoral musculature integrated by the weakly developed DPC of the humerus (char. 138.2) (Pei et al., 2020). With that in mind, we can really picture the fact that the scansorial form of locomotion and gliding form of flight found anchiornithines to be the possible starting point of how true avian powered flight originated. However, the specialized gliding found in scansoriopterygids and gliding in anchiornithines differentiate from one another and are thus considered among different methods of flight. It is not to say the anchiornithines may be ancestral to the point where we think that is how flight started. Instead, we could think of the anchiornithines as a way to introduce characteristics for true powered flight and rather treat them as part of a very large evolutionary experiment. A trial-and-error experiment that originated at the root of the family tree of Pennaraptoran theropods that would give way to modern birds and modern powered flight through the process of convergent evolution.

Comparisons between *Archaeopteryx* and *Anchiornis*

Among the anchiornithine group of theropods, there is *Anchiornis huxleyi*. Its recent addition to the fossil record gives us a more complex picture of how avian flight evolved among the theropods. In fact, there are a ton of comparisons to that of *Archaeopteryx* we must immediately address. We shall begin with the sternum and the pectoral girdle. The sternum in *Anchironis*, whether it be chondrified or ossified, is absent as in other basal avialans such as *Archaeopteryx* and *Sapeornis* (Zheng et al., 2014). It just so happens that the absence of the sternum is one of the morphological similarities uniting *Archaeopteryx* and anchiornithine theropods (e.g., *Anchiornis* and *Xiaotingia*) as well as other postcranial skeletal morphologies such as a short, quadrangular coracoid, a triradiate ischium, and a gastral basket composed of 12-14 pairs of gastralia (Hu et al., 2009; Zheng et al., 2014). This not only show signs of morphological convergence between *Archaeopteryx* and *Anchiornis* (seeing that *Anchiornis* and *Xiaotingia* and others are recently placed near the base of Avialae (Pei et al., 2020)), but they also show signs of morphological constraints that prohibit powered flapping flight. Furthermore, because of these constraints, aerial locomotion may have started on the ground with some occasion of WAIR (wing-assisted inclined running) or other lesser means of aerial locomotion before the development of more, veined asymmetrical feathering I more

derived paravian theropods (e.g., *Microraptor*), basal avialans (e.g., *Confuciusornis*), and modern birds. In addition to the absence of the sternum, *Archaeopteryx* and paravian theropods, such as *Anchiornis*, share a similar morphology of the pectoral girdle with the rhea, which is that the pectoral girdle has glenoids laterally oriented and with the major axis subvertical (Agnolin et al., 2019; Novas et.al., 2020). In early-diverging birds (e.g., *Archaeopteryx, Anchiornis*), the glenoid facet retained the lateral orientation seen in other early-diverging proavians (e.g., *Buitreraptor*), and with both sections (scapular and coracoidal) subequal in size and aligned on the same dorsoventral plane (Jenkins; 1993; Baier et al., 2007; Novas et al., 2018; Agnolin et al., 2019). The lack of a "twisted" glenoid (that is, without a large coracoidal surface dorsally faced) suggests that early-diverging birds (e.g., *Archaeopteryx, Anchiornis*) did not attain the amplitude of forelimb movements, nor the humeral trajectory over the glenoid, as characteristically occurs in living flying avians (Ostrom, 1976; Poore et al, 1997). The general resemblance of pectoral girdle and humerus of early-diverging theropods (e.g., *Buitreraptor, Bambiraptor,*

Deinonychus (Ostrom, 1969; Burhnam, 2004)) preserved three-dimensionally on slabs suggest that the latter one retained almost the same forelimb posture and range of movements as in other early-diverging paravians. *Archaeopteryx* and *Anchiornis* both have pectoral girdles similar to the rhea and are below the standards of the pectoral girdle of modern flying birds. Therefore, these similarities in the pectoral girdle indicate an ancestral condition sometime during the Late Jurassic Period. However, despite both of those animals having similarities on the pectoral girdle, the difference lies on the outside. I am talking about the feathers.

Pan et. al. (2019) investigated the molecular composition of the feathers of *Anchiornis* to see if *Anchiornis* was capable of flight (for real) despite the presence of pennaceous feathers. Results from molecular and ultrastructural analysis showed that the feathers of *Anchiornis* were made of both short and thin (3 mm.) filaments consisted of – keratin and a greater number of thick filaments with a diameter of about 8 mm., making them consistent with the basal – keratins (Pan et al., 2019). This indicates that the feathers of *Anchiornis* expressed both – keratins and – keratins, which are found in both embryonic and mature feathers of extant birds (Wu et al., 2015). The study also found that because the mature feathers of extant birds are dominant by the feather – keratins, the coexpression of both – and feather – keratins, in combination with the ultrastructural patterns, show that feathers of *Anchiornis* would most likely represent an evolutionary transition between the ancestral integumentary appendages and extant bird feathers (Pan et al., 2019).

Based on Pan et al's study (2019) of feathers in early theropods, we could deduce that the feathers also show support of the fact that paravian theropods like *Anchironis, Serikornis,* and other anchironithine paravian theropods may not have flown well as *Archaeopteryx* and/or more derived birds like *Confuciusonris, Jeholornis*, and modern birds. In fact, the pennaceous feathers of *Anchiornis* are likely an intermediate stage of feather evolution as they may have lacked the biomechanical properties suitable for flight as suggested by the molecular composition and ultrastructure of the feathers in *Anchiornis* (Pan et al., 2019). So, it is possible that since *Anchiornis* had this mixture of feather – and – keratins, the "shaggy" appearance (Saitta et al., 2018), absence of the sternum, and symmetry of the pennaceous feathers in the fore- and hindlimbs, it would have also posed as an intermediate or even primitive stage of flight. Maybe by adding all of this together, it would make more sense that flight was neither powered nor advanced in comparison to *Microraptor* and/or modern birds as far as the anchiornithines. Flight could have started out imperfect when they, along with *Archaeopteryx* originated in the Late Jurassic as this moment in Earth's history was the start of multiple forms of variation in both evo-developmental and morphological sense.

Sources

Agnolin, F. L., Motta, M. J., Brissón Egli, F., Lo Coco, G., & Novas, F. E. (2019). Paravian phylogeny and the dinosaur-bird transition: an overview. *Frontiers in Earth Science*, *6*, 252.

Baier, D. B., Gatesy, S. M., & Jenkins, F. A. (2007). A critical ligamentous mechanism in the evolution of avian flight. *Nature*, *445*(7125), 307-310.

Burnham, D.A. 2004. New Information on *Bambiraptor feinbergi* from the Late Cretaceous Montana. *In* P.J. Currie, E.B. Koppelhus, M.A. Shugar, and J.L. Wright (editors), Feathered Dragons: Studies on the Transition from Dinosaurs to Birds: 67-111. Indianapolis: Indiana University Press.

Byrnes, G., Libby, T., Lim, N. T. L., & Spence, A. J. (2011). Gliding saves time but not energy in Malayan colugos. *Journal of Experimental Biology*, *214*(16), 2690-2696.

Cheney, J. A., Allen, J. J., & Swartz, S. M. (2017). Diversity in the organization of elastin bundles and intramembranous muscles in bat wings. *Journal of Anatomy*, *230*(4), 510-523.

Czerkas, S. A., & Feduccia, A. (2014). Jurassic archosaur is a non-dinosaurian bird. *Journal of Ornithology*, *155*(4), 841-851.

Dececchi, T. A., Roy, A., Pittman, M., Kaye, T. G., Xu, X., Habib, M. B., ... & Zheng, X. (2020). Aerodynamics show membrane-winged theropods were a poor gliding dead-end. *Iscience*, *23*(12), 101574.

Feduccia, A. (2012). Chapter 4: Mesozoic Chinese Aviary Takes Form. In *Riddle of the Feathered Dragons: Hidden birds of China* (p. 154). essay, Yale University Press.

Foth, C., & Rauhut, O. W. (2017). Re-evaluation of the Haarlem Archaeopteryx and the radiation of maniraptoran theropod dinosaurs. *BMC Evolutionary Biology*, *17*(1), 1-16.

Heilmann, G. (1926). The Proavian. In *The Origin of Birds* (pp. 193–201). essay, H.F. & G. Witherby.

Hwang, S. H., NORELL, M. A., Qiang, J. I., & Keqin, G. A. O. (2002). New specimens of Microraptor zhaoianus (Theropoda: Dromaeosauridae) from northeastern China. *American Museum Novitates*, *2002*(3381), 1-44.

Hu, D., Clarke, J. A., Eliason, C. M., Qiu, R., Li, Q., Shawkey, M. D., ... & Xu, X. (2018). A bony-crested Jurassic dinosaur with evidence of iridescent plumage highlights complexity in early paravian evolution. *Nature Communications, 9*(1), 1-12.

Hu, D., Hou, L., Zhang, L., & Xu, X. (2009). A pre-Archaeopteryx troodontid theropod from China with long feathers on the metatarsus. *Nature, 461*(7264), 640-643.

Jenkins, F. A. (1993). The evolution of the avian shoulder joint. *American Journal of Science, 293*(A), 253-267.

Kellner, A. W., Wang, X., Tischlinger, H., de Almeida Campos, D., Hone, D. W., & Meng, X. (2010). The soft tissue of Jeholopterus (Pterosauria, Anurognathidae, Batrachognathinae) and the structure of the pterosaur wing membrane. *Proceedings of the Royal Society B: Biological Sciences, 277*(1679), 321-329.

Lefevre, U., Cau, A., Cincotta, A., Hu, D., Chinsamy, A., Escuillié, F., & Godefroit, P. (2017). A new Jurassic theropod from China documents a transitional step in the macrostructure of feathers. *The Science of Nature, 104*(9), 1-13.

McGuire, J. A., & Dudley, R. (2011). The biology of gliding in flying lizards (genus Draco) and their fossil and extant analogs. *Integrative and Comparative Biology, 51*(6), 983-990.

Motta, M. J., Agnolín, F. L., Brissón Egli, F., & Novas, F. E. (2020). New theropod dinosaur from the Upper Cretaceous of Patagonia sheds light on the paravian radiation in Gondwana. *The Science of Nature, 107*(3), 1-8.

Novas, F. E., Egli, F. B., Agnolin, F. L., Gianechini, F. A., & Cerda, I. (2018). Postcranial osteology of a new specimen of Buitreraptor gonzalezorum (Theropoda, Unenlagiidae). *Cretaceous Research, 83*, 127-167.

Ostrom, J. (1976). Some hypothetical anatomical stages in the evolution of avian flight. Smithsonian Contributions to Paleobiology, 27: 1–21.

Ostrom, J. H., & Gauthier, J. A. (2019). *Osteology of Deinonychus antirrhopus, an unusual theropod from the Lower Cretaceous of Montana*. Yale University Press.

Palmer, C. (2018). Inferring the properties of the pterosaur wing membrane. *Geological Society, London, Special Publications, 455*(1), 57-68.

Pan, Y., Zheng, W., Sawyer, R. H., Pennington, M. W., Zheng, X., Wang, X., ... & Schweitzer, M. H. (2019). The molecular evolution of feathers with direct evidence from fossils. *Proceedings of the National Academy of Sciences, 116*(8), 3018-3023.

Pei, R., Li, Q., Meng, Q., Gao, K. Q., & Norell, M. A. (2014). A new specimen of Microraptor (Theropoda: Dromaeosauridae) from the Lower Cretaceous of western Liaoning, China.*American Museum Novitates*,*2014*(3821), 1-28.

Pei, R., Pittman, M., Goloboff, P. A., Dececchi, T. A., Habib, M. B., Kaye, T. G., ... & Xu, X. (2020). Powered flight potential approached by wide range of close avian relatives but achieved selectively.*BioRxiv*.

Poore, S. O., Sanchez-Haiman, A., & Goslow, G. E. (1997). Wing upstroke and the evolution of flapping flight.*Nature*,*387*(6635), 799-802.

Saitta, E. T., Gelernter, R., & Vinther, J. (2018). Additional information on the primitive contour and wing feathering of paravian dinosaurs.*Palaeontology*,*61*(2), 273-288.

Socha, J. J., Jafari, F., Munk, Y., & Byrnes, G. (2015). How animals glide: from trajectory to morphology.*Canadian Journal of Zoology*,*93*(12), 901-924.

Sorkin, B., & Somerset, K. Y. Sorkin, B. 2020. Scansorial and aerial ability in Scansoriopterygidae and basal Oviraptorosauria. Historical Biology.

Wang, M., O'Connor, J. K., Xu, X., & Zhou, Z. (2019). A new Jurassic scansoriopterygid and the loss of membranous wings in theropod dinosaurs.*Nature*,*569*(7755), 256-259.

Wang, X., Pittman, M., Zheng, X., Kaye, T. G., Falk, A. R., Hartman, S. A., & Xu, X. (2017). Basal paravian functional anatomy illuminated by high-detail body outline.*Nature communications*,*8*(1), 1-6.

Wu, P., Ng, C. S., Yan, J., Lai, Y. C., Chen, C. K., Lai, Y. T., ... & Chuong, C. M. (2015). Topographical mapping of α-and β-keratins on developing chicken skin integuments: Functional interaction and evolutionary perspectives. *Proceedings of the National Academy of Sciences*,*112*(49), E6770-E6779.

Xu, X., Zhou, Z., Wang, X., Kuang, X., Zhang, F., & Du, X. (2003). Four-winged dinosaurs from China.*Nature*,*421*(6921), 335-340.

Yang, Z., Jiang, B., McNamara, M. E., Kearns, S. L., Pittman, M., Kaye, T. G., ... & Benton, M. J. (2019). Pterosaur integumentary structures with complex feather-like branching.*Nature ecology & evolution*,*3*(1), 24-30.

Zheng, X., O'Connor, J., Wang, X., Wang, M., Zhang, X., & Zhou, Z. (2014). On the absence of sternal elements in Anchiornis (Paraves) and Sapeornis (Aves) and the complex early evolution of the avian sternum.*Proceedings of the National Academy of Sciences*,*111*(38), 13900-13905.

https://terribleliz ards.libsyn.com/ website

(210) 494-4507

Welcome to the Podcast!

SKORPIOVENATOR REAL OR MADE UP DINOSAUR

BY DR. BRIAN CURTICE

SKORPIOVENATOR

Is *Skorpiovenator* a real dinosaur?

Skorpiovenator was named in... 2008 and it is most definitely a real dinosaur. This incredible Late Cretaceous Argentine abelisaurid is known from a spectacularly preserved skeleton missing only parts of the tail and its arms. The name means "scorpion hunter," so named "...because of the abundance of living scorpions moving around the excavation". Yikes!

At around 20' long and over 3,000 lbs, *Skorpiovenator* was a medium-sized abelisaur. The largest currently known abelisaurid is *Pycnonemosaurus* at 30' long and considerably heavier.

The abelisaurs, that we know of, never attained the massive size that tyrannosaurids and carcharodontosaurids attained. Yet, in the southern hemisphere, the carcharodontosaurids go extinct around 90 million years ago, while the abelisaurs make it right up to "The Last Day".

Abelisaurs are wildly successful Gondwanan theropods known from all over the southern hemisphere with new taxa being named every hear. *Skorpiovenator* lived alongside the theropods *Mapusaurus, Taurovenator, Ilokelesia, Gualicho, Tralkasaurus, Aoniraptor, Ovroraptor, Huinculsaurus*, the sauropods *Argentinosaurus, Choconsaurus, Cathartesaura*, and *Limaysaurus*, and unnamed iguanodonts.

Abelisaurids are famous for two amazing features, their bizarre arms and their crazy cool skulls. How much of the *Skorpiovenator* arm complex exists? The field photo and available cast do not show any evidence of a shoulder blade, nor any of the arm bones themselves. The descriptive text says, "...lacking the right forearm...", implying the left arm is present, however, the accompanying skeletal illustration shows only one bone drawn in the left arm, either the radius or the ulna. Character 88, "Forelimb zeugopodium composed by short and robust radius and ulna", is marked as a '1', meaning at least one of the forearm bones is present. However, the character reads "radius and ulna", implying both are present, which is not supported by the illustration. Character 90, "Distal end of radius/ulna...", is marked with a question mark, suggesting whatever forelimb bone(s) collected are either missing the distal half or were not prepared for the study. There is no other mention of the forelimb in the text. No subsequent papers have mentioned its arms either. The missing components of the tail were touched upon in the original description but for reasons unknown, nothing else was said about the shoulder or arms. Though I hope the left arm is complete and simply needs preparation I am of the opinion it doesn't exist other than a fragment of a forelimb bone (Tracy Ford's Paleofiles lists it as an ulna). One of the best abelisaurid forelimbs discovered thus far is that of Majungasaurus. It has a gigantic shoulder blade for such little hands. The four fingers, if one can call them that, consist of metacarpals (the wide, flat part of your hand), then a 1-2-1-1 phalanx count. This means it had only 1 "finger bone" (the first bone you can see at the base of your finger), then 2 bones on the next finger over, then the remaining fingers had only one finger bone each. It may not even have had claws! What the heck they were doing with stumps I have no idea. The entire arm was short, making T. rex arms seem kingly. Maybe they slap-fought for dominance? Were they somehow used for display to attract mates? Skorpiovenator's completeness led me to hope an arm was present but, alas, it seems not to be the case.

Skorpiovenator's skull doesn't have the horns of *Carnotaurus*, however it does have a very rough face, with lots of texture. It is possible the skull was covered in thick, extremely robust , bumpy scales. The scales may have even been like armor, they were that thick!

The skull photo in the original publication reminded me of the marine iguanas (*Amblyrhynchus cristatus*) I studied in the Galapagos. On numerous occasions, I observed them fight one another. They pushed their bumpy heads against one another, the long toes and toe claws grasping for purchase on the sand and rocks. I marveled at one battle where the smaller competitor, to me, won the match for it was able to push the larger one around using better technique (or perhaps just was lucky with the substrate). The larger male at one point was rolled over and took the opportunity to bite the smaller male. The head butting pushed the larger male's head repeatedly into rocks, resulting in a bloody lip. Yet, for reasons I know not, the fight simply stopped. Ah lizard culture Perhaps*Skorpiovenator*used its skull in a similar fashion, slow-speed headbutting that relied upon its powerful legs (those tiny arms wouldn't be able to do anything!) and rigid neck region to push an opponent back or even knock it down where it could kick and bite if the loser didn't acquiesce.

Or, perhaps they used their heads to side-strike opponents like giraffes do today. The bumpy-textured skull was robustly built, keeping the brain safe, and the neck bones are quite rigid, giving it. Such bashing would be quite the sight to see!

Abelisaurid tails have some of the most beautiful caudal ribs one will ever see! In fact, the way they are built, it is quite likely the base of the tail was quite inflexible. This tracks with modifications in the dorsal and cervical vertebrae, suggesting the vertebral column was rod-like and fairly inflexible.

Watch more here!

https://www.fossilcrates.com/blogs

Skorpiovenator bustingorryi

Kingdom: Animalia
Phylum: Chordata
Clade: Dinosauria
Clade: Saurischia
Clade: Theropoda
Family:† Abelisauridae
Clade:† Brachyrostra
Genus:† Skorpiovenator
Type species†
Skorpiovenator bustingorryi

https://www.geologyflannelcast.com/

the GEOLOGY FLANNEL-CAST

HOW PLESIOSAURS SWAM: NEW INSIGHTS INTO THEIR UNDERWATER FLIGHT USING "AVA", A VIRTUAL PLIOSAUR

BY MAX HAWTHORNE[1], *MARK A. S. MCMENAMIN[2], PAUL DE LA SALLE[3]

[1] Far From The Tree Press, LLC, 4657 York Rd., #952, Buckingham, PA, 18912, United States
[2] Department of Geology and Geography, Mount Holyoke College, South Hadley, Massachusetts, United States
[3] Swindon, England

*Correspondence: author@maxhawthorne.com; Tel.: 267-337-7545

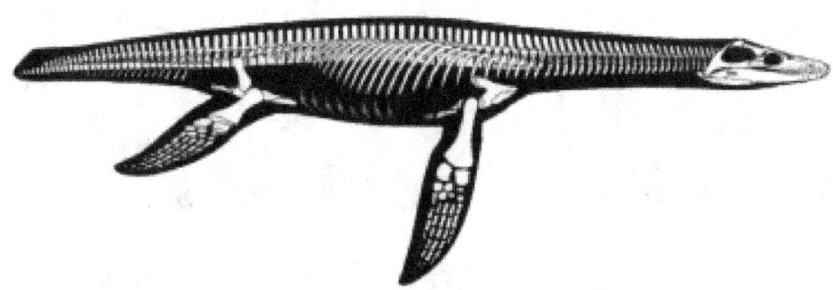

Figure 4. *Rhomaleosaurus*, based on Smith/Benson 2014 but with adjusted neutral (resting) positioning for the paddles. (used with permission).

Abstract Analysis of plesiosaur swim dynamics by means of a digital 3D armature (wireframe "skeleton") of a pliosauromorph ("Ava") demonstrates that: 1, plesiosaurs used all four flippers for primary propulsion; 2, plesiosaurs utilized all four flippers simultaneously; 3, respective pairs of flippers of Plesiosauridae, front and rear, traveled through distinctive, separate planes of motion, and; 4, the ability to utilize all four paddles simultaneously allowed these largely predatory marine reptiles to achieve a significant increase in acceleration and speed, which, in turn, contributed to their sustained dominance during the Mesozoic.

Keywords aquatic reptiles, plesiosaurs, pliosaurs, swim kinematics, Strouhal numbers

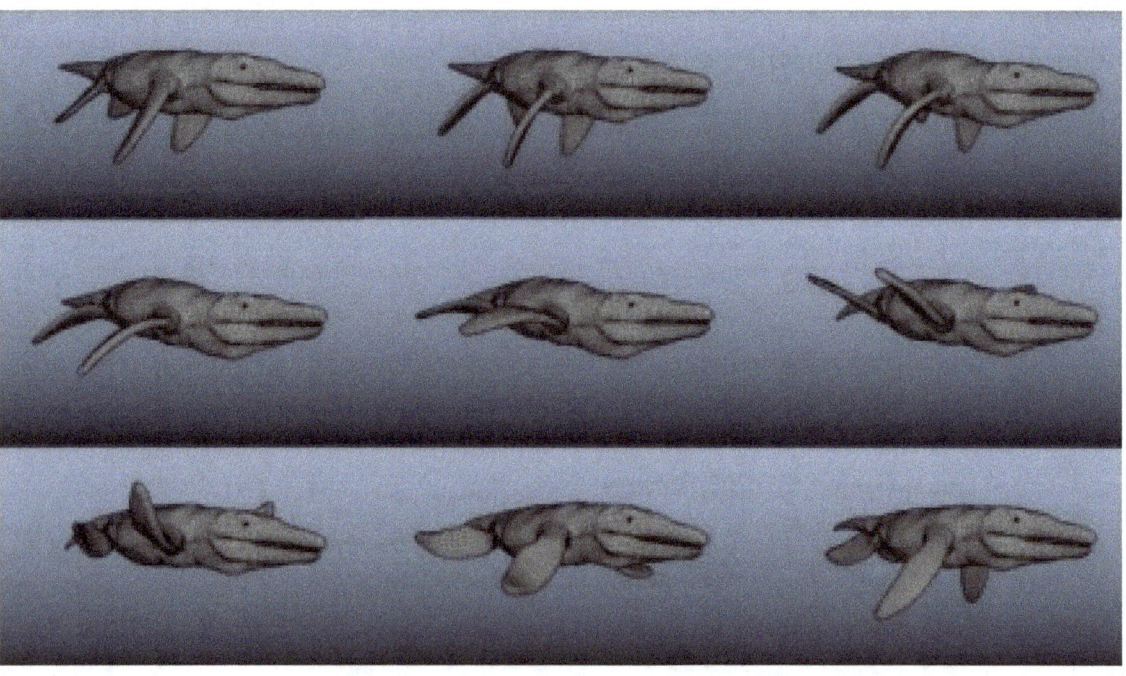

Figure 13. Wireframe swim cycle of Ava (animation by Mathieu Lafreniere).

1. Introduction

Plesiosauridae are a clade of extinct Mesozoic marine reptiles with a long evolutionary history. Since their discovery in 1821, the precise method of their swim locomotion has remained unsolved, although many hypotheses have been proposed [1-6]. Previous studies have posited a variety of locomotory techniques to offset the implied redundancy of the paired paddles working in conjunction while moving through the same subaqueous plane; i.e. a duplication of resistance that generated little or no additional thrust (herein referred to as the "principal of flipper redundancy"). The only thing certain is that plesiosaurs had abandoned the tail propulsion common in earlier aquatic reptiles [7]. Mosasaurs and ichthyosaurs, in contrast, continued with tail propulsion until the Cretaceous [8]

Figure 1. *Peloneustes* sp. femur, showing muscle attachment site. Specimen from the de la Salle collection.

Modelers have proposed front only (or front primary) means of propulsion, a rear only (or rear primary) means of propulsion, an alternate paddle application means of propulsion, and a figure-eight pattern paddle application [9]. Further study attempted to justify the use of all four flippers simultaneously via the use of paddle-generated vortices, which require specific timing to achieve optimal additional thrust. These attempts have largely relied on anatomical studies of strata-compressed plesiosaur skeletons, and/or preconceived notions as pertains to the paddles' inherent ranges of motion [8, 10-12]. What has not been considered are the opposing angles of the pectoral and pelvic girdles, which strongly indicate varied-yet-complementing relations between the front and rear sets of paddles, both in repose and in motion, and imply separate planes of motion for both. The main argument when it comes to defining plesiosaur locomotion has been centered on the notion that two sets of flippers cannot be effectively deployed simultaneously for an effective thrust stroke. If they were, they would be pushing the same water, hence the animal couldn't go significantly faster. As per the principle of flipper redundancy, why use four fins to do a job when two are sufficient, as in sea lions? Many hypotheses have been put forward to explain why plesiosaurs had four flippers and also to explain how they used them. These included: the use of alternate strokes (front

then rear) which enabled one set of flippers to rest while the other worked; the use of rear or front flippers as a primary means of propulsion (with the remaining set delegated mainly to steering), and; the "vortices theory", where paddle-generated vortices assist the animal's recovery stroke. Do any of these previous hypotheses explain plesiosaur locomotion? If not, how did the four paddles function?

Based on an overview of known species, we believe the concept of an alternate flipper or figure-eight pattern is mistaken. If we consider modern marine life: fish, cetaceans, or the usual list of tetrapods compared to plesiosaurs, due to their reliance on a similar, albeit single set of flippers for propulsion (penguins, pinnipeds and sea turtles), we see that a single caudal fin or set of paddles is sufficient. There is no advantage for a duplicate or redundant set of flippers, i.e. no extant species uses two sets of equally-developed flippers with one pair that must "rest" during its negative stroke in order to recover. Therefore, there must be a locomotory advantage for the development of four, relatively equal paddles.

Based on available evidence, we believe that the notion of a front or rear-only swim pattern is even less plausible. When examining the skeletons of plesiosaurs, we posit here a unique radial adaptation of the shoulder and a similar mechanism developed in the pelvis. The bones for both pairs of flippers are extremely well-developed with large muscle attachment points (Figure 1). The muscles powering these paddles were undoubtedly huge and made for strength and endurance.

Modern examples provide some context for our hypothesis. Consider the recently discovered waterfall-climbing blind cave fish, Cryptotora thamicola [13]. It has developed a pelvis and a stiffened spine, giving it the ability to utilize four comparably developed fins for locomotion. It uses all of them simultaneously, with a tetrapod-like gait that resembles that of a salamander. Consider also the case of extant sea turtles (Fig. 2). The large,

Figure 1. *Peloneustes* sp. femur, showing muscle attachment site. Specimen from the de la Salle collection.

well-developed forelimbs are the propulsion units and are also responsible for major directional changes, whereas the smaller, more restricted rear flippers function mainly as rudders to keep the animal on an even keel, assist in turning, and, in the case of females, for nest excavation. Two swimming gaits have been characterized in turtles [14]. A similar configuration of smaller rear flippers was employed by the parvipelvian ichthyosaurs. Small flippers are quite adequate for steering purposes.

The hypothesis that a plesiosaur's rear flippers benefited from dual lines of vortices (referred to as the "vortex street") generated by the front set depends on both sets of paddles moving through the same planes of motion, as well as at a certain pace and spacing, in order to effectively generate additional thrust [6]. Both marine and terrestrial tetrapods rely on a varying gait based on circumstances (changing direction, accelerating, etc.), and would be ill-equipped to survive were they limited to a unique, "optimal" speed/phasing in order to derive maximum benefits. The robot used to test the hypothesis [6] was equipped with a paddle that allows rotational freedom of movement.

We propose that skeletal adaptations in plesiosaurs enabled them to utilize all four flippers simultaneously, and in such a way that the paddles not only avoided violating the principal of flipper redundancy but also complemented one another by moving in conjunction through distinctly separate planes of motion. By doing so, this adaptation enabled the animal to generate significantly more thrust, increasing both its ability to accelerate as well as its maximum speed.

Range of motion limitations indicated by studying non-compressed skeletons of plesiosaurs suggest that the front flippers pushed down and back, in a pattern similar to that of a penguin, while the rear flippers pushed more horizontally and backwards. The front flippers therefore moved through a lower plane of motion than the rear, thus pushing water in a different range. When working in combination they moved twice as much water, giving the animal distinctive and unique advantages, comparable to what one might see in a fish or whale with a tail twice the original size or a sea lion with two sets of flippers.

2. Methodology

For purposes of this study, a multi-pronged approach was adopted. This included comparison studies of physical and photographic evidence of the skeletal remains of an assortment of known Plesiosauridae forms, including plesiosauromorph and pliosauromorph body types, as well as polycotylids. Comparative references were also made with extant tetrapods (Cheloniidae and Spheniscidae) which utilize front-paddle-only, dual-paddle locomotory methods similar to plesiosaurs. Data acquired from the aforementioned anatomical studies was utilized in the development of a computer model which, taking water resistance into account, went through various iterations until it successfully recreated a functional plesiosaur swim cycle, one that mimics the underwater movements of these remarkable marine Sauropsida.

As previously stated, traditional views and the subsequent positioning of plesiosaur paddles in attempted motion reconstructions have been largely based on taphonomically flattened remains and/or the assumption that the flippers moved in a restricted range of motion (Fig. 3). This applies to almost all previously published theories [6, 15]. However, when observing a more scientifically accurate, non-compressed skeleton (Figs. 3-4), it becomes obvious that there are inherent differences in the respective ranges of motion of the front and rear paddles, as evidenced by

morphological adaptations of the skeleton, in particular, at the humeral and femoral insertions. The elongate, blade-like dorsal process of the clavicle of the polycotylid plesiosaur Manemergus anguirostris from the Turonian of Morocco [16], for example, indicates that the third dimension is critial for understanding muscle attachments and range-of-motion. When observing the fore-flippers' humeral insertion point (Figure 4), it is evident that the pectoral/shoulder girdle angles upward, toward the animal's cranium. This elevation of the shoulder girdle implies that, when generating thrust, the front paddles were able to angle upward and sweep

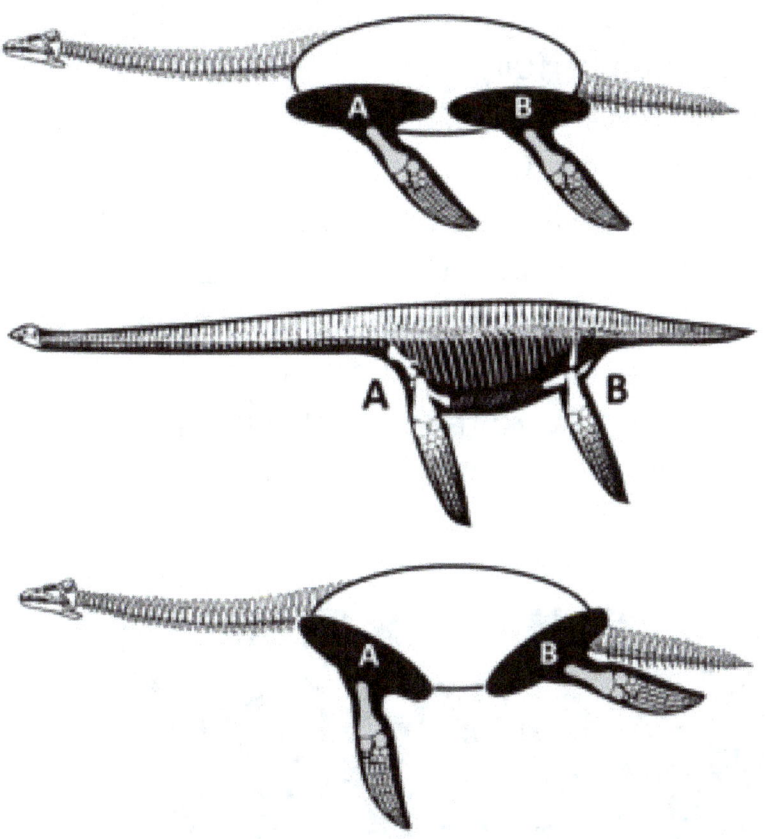

Figure 3. Adjustments to the positioning of front (A) and rear (B) plesiosaur paddles. Adapted from *Plesiosaurs Neo* by Professor Ryosuke Motani and www.plesiosauria.com (Brown 1981, used with permission).

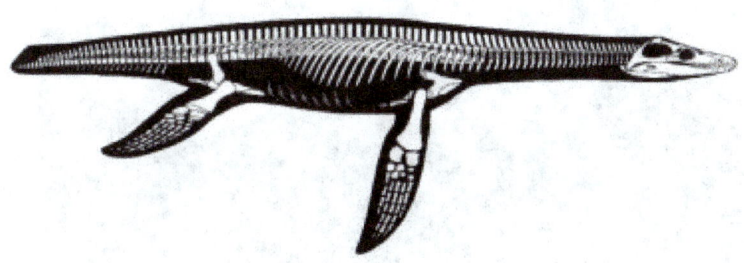

Figure 4. *Rhomaleosaurus*, based on Smith/Benson 2014 but with adjusted neutral (resting) positioning for the paddles. (used with permission).

further forward on the negative stroke that the rear set, coming back and down powerfully on the positive. The power-stroke from the front limbs/paddles traveled through a largely vertical movement, not unlike that of a penguin, except with a bit more range and flexibility, and most likely terminated in the same direction shown in Figure 5.

When observing the femoral insertion point where the rear flippers insert into the pelvic girdle, we find that the pelvis is angled in the reverse – up toward the caudal vertebrae (Figure 4). This change in angle/range of motion physically

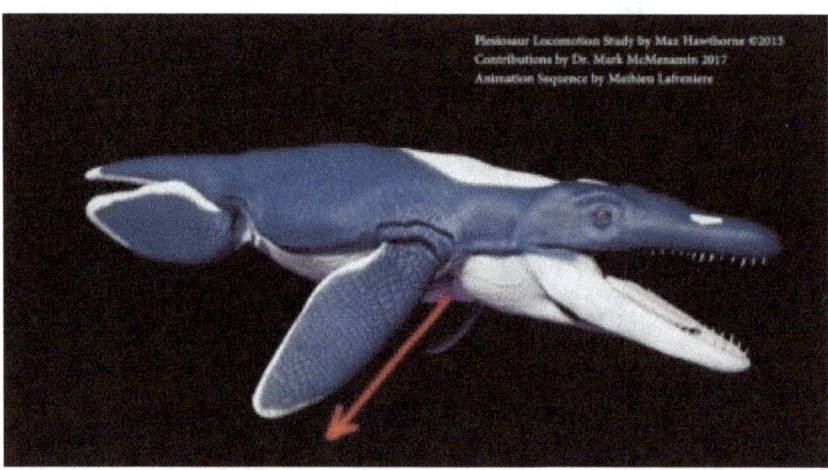

Figure 5. Downstroke of the plesiosaur fore flipper.

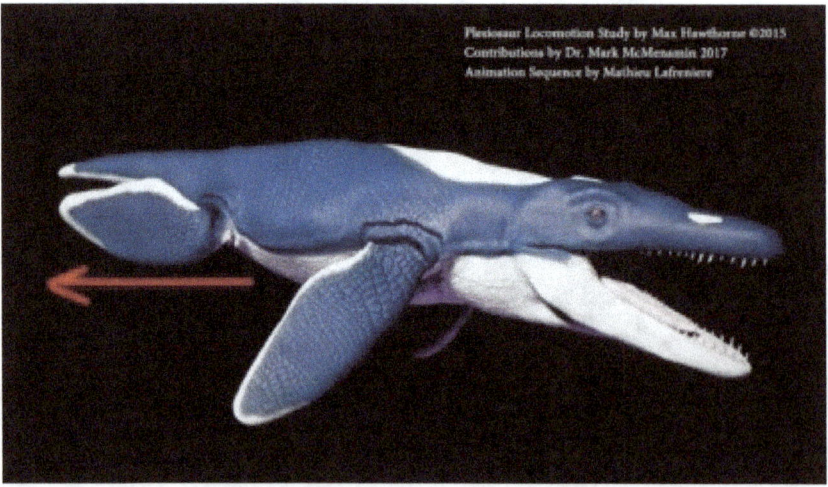

Figure 6. Backward thrust of the plesiosaur hind flipper.

restricted the animal's ability to thrust down hard during a running speed downstroke, but, at the same time, it enabled it to raise its rear set of flippers higher in the water throughout its stroke, pushing backward and ending in a far more lateral termination point than the front set. This served the purpose of encompassing a largely different plane of motion, enabling the plesiosaur to thrust back hard with its posterior paddles, most likely finishing in a position more aligned with its body (Figure 6).

Strouhal Number at Cruising Speed

The main goal of our digital animation model, Ava (a generalized brachauchenine thalassophonean pliosaurid close to Kronosaurus boyacensis), is to assess the swimming mechanics of an ancient marine predator. Before proceeding to the results of the simulation, we report here a simple goodness-of-fit test for the model to determine if it generates plausible metrics in terms of cruising speed swim using some reasonable assumptions.

This involves a calculation of a dimensionless number, the Strouhal value for Ava's cruising speed. Strouhal number (St), a measure of the efficiency of propulsion, is calculated as:

$$St = fA/U$$

where f equals stroke frequency, A equals stroke amplitude, and U equals forward speed. Assuming a pliosaur 11 m length, we estimate a cruising speed stroke frequency (f) of 0.5 beats/sec, a stroke amplitude of 1.48 m, and a cruising speed of 3 mps.

A frequency of one stroke every two seconds is identical to that of a cruising leatherback (Dermochelys corinacea). Using the above values gives Ava's Strouhal number as:

$$St_A = 0.247.$$

Strouhal number peaks in living aquatic tetrapod lineages at between $0.2 < St < 0.4$. This has been considered a case of tuning for higher power efficiency [17]. At approximately 2.5, St_A is well within this range, showing that the Ava model generates a reasonable Strouhal value when calculated with reasonable stroke frequency and speed assumptions. Stroke amplitude was directly measured from a still shot of the animation, and was measured without consultation with or

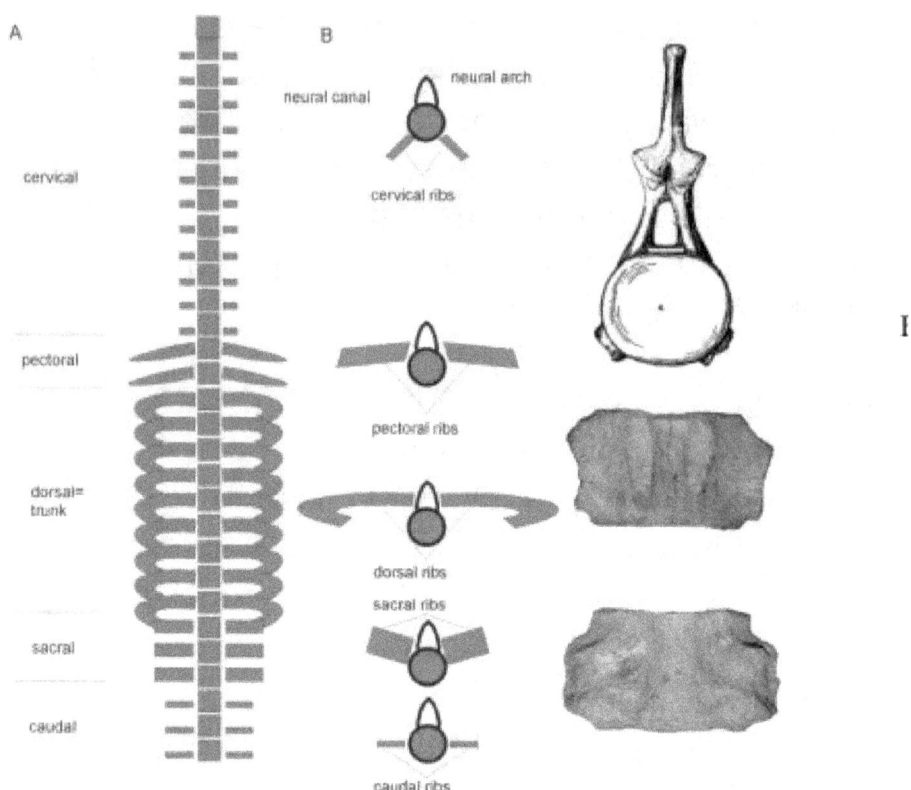

Figure 7. Plesiosaur vertebral column diagram (adapted from O'Gorman/Fernandez 2016, used with permission)

knowledge of our animator Mathieu Lafreniere, and so was largely independent of rendering the animation itself. StA is in the lower half of the Strouhal peak efficiency zone, but this might not be unexpected for a predatory animal that might be expected to have optimized its burst or running speed at the expense of its cruising speed. We do not attempt here to calculate Strouhal value for burst speed, as burst speed estimates are poorly constrained at this time (see below).

Comparative Swim Kinematics

Evolutionary adaptations to the rib cages of plesiosaurs support our hypothesis, that is, simultaneous use of all four flippers. Both the pectoral and sacral ribs observed in many plesiosaurs are characteristically shortened, the sacral ribs above the pelvic girdle typically more so than those above the pectoral (Figures 4, 7

As in sea turtles, which often have an "alcove" in their shells to accommodate their front flippers upward lift (Figure 8), morphological adaptations typically reduce friction and impacts by moving body parts. Turtle flippers have large, powerful muscles, and as they are flexing and hoisting the flippers up and back both those muscles, and the flippers themselves, would rub against ribs that extended too far down. Hence, they became reduced in length. In plesiosaurs, the amount of adaptive rib shortening that occurs is consistent with our theorized range of motion for the respective flipper pairs. We see standard, esophageal-accommodating shortening in the cervical ribs, moderate, but more pronounced, pectoral rib shortening for the front set of paddles and extreme shortening, combined with upward angling, of the sacral ribs, to accommodate both the higher range of motion of the rear paddles and the flexion of the muscles driving them (Figures 7, 9).

Figure 8. *Chelonia mydas*, the green sea turtle (licensed image).

This pattern of adaptive rib shortening occurs repeatedly in assorted plesiosaur species, with homoplasy occurring in the angles of the front and rear limb girdles, with the greatest degree of shortening taking place in the pelvic/sacral region (Figures 7-9). This allowed the animal's front limbs to move in a pattern somewhat akin to a penguin or sea turtle (sans the chelonian's full upper range, due to the absence of an elbow joint), with the rear flippers taking advantage of their ability to reach higher and encompassing water the front flippers were unable catch. During the return, reverse, or negative stroke, the movement of the two sets would have been similar. It is likely that, during high speed pursuits, the rear flippers benefited from the suction/vacuum generated by the front flippers during their reverse stroke. The rear pair could practically come along for the ride, conserving their strength and absorbing kinetic energy to allow for a more powerful positive stroke.

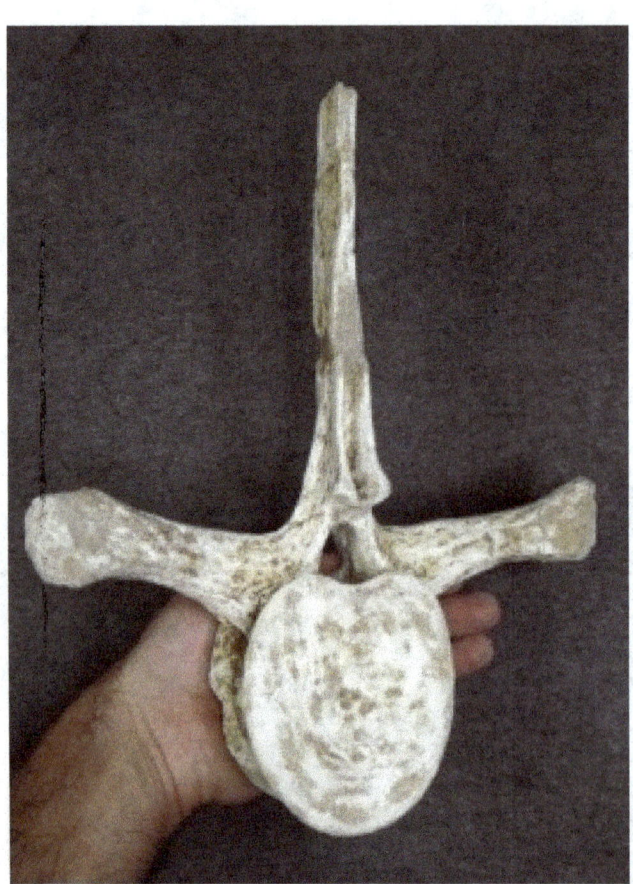

Figure 9. *Zarafasaura oceanis* sacral (from the Hawthorne collection).

This pattern of movement is supported by Frank Sanders' and Kenneth Carpenter's 2010 study [18], which concluded that the flipper movement in plesiosaurs was largely vertical: "The best modern anatomical analogs are sea turtles and underwater birds, particularly penguins. All of these animals swim with variations of underwater flight. In particular, they all move their limbs mainly vertically, with some fore-to-af motion and slight axial rotation, all of which is possible for the plesiosaur limbs."

Consideration of this largely vertical movement, but taking into account the adjusted ranges of motion (implied by the opposingly angled pectoral and pelvic girdles of plesiosaurs), demonstrates how these animals were able to utilize both sets of paddles – front and rear – moving them through separate planes or motion, and to maximum benefit. Using the above joint motion constraints and the resultant theoretical ranges of motion as a baseline, and taking into account the separate planes of motion for both the front and rear paddles, a digital 3D armature (wireframe "skeleton") of a pliosauromorph (known as "Ava") was then constructed using the Autodesk Maya 3D Computer Graphics Application. Assorted ranges of motion and combinations of flipper movements and angles were utilized in the experiment, including the incorporation of various rhythms and speeds.

Once the "flesh" was added to Ava's armature, we studied footage of extant penguins, sea lions, and sea turtles to help us understand and take into account directional changes that took place based on the angles of each thrust, as well as the effects said thrust had on the animal's body. Issues of stability were addressed, with the final model/sequence goal to develop a virtual plesiosaur that could perform a perfected swim cycle at various speeds, including rapid acceleration and gliding.

and gliding.

3. Results and Discussion

As the accompanying video demonstrates, the results of our study indicated that plesiosaurs did indeed utilize both sets of paddles for locomotion. Moreover, they used them simultaneously, with the flippers traveling through separate planes of motion to not only maximize benefits in terms of generating initial thrust, but also to achieve higher top end speed as well as increased maneuverability. Over the course of our study our virtual pliosaur Ava went through numerous revisions. Each iteration took into account studies of the skeletal remains of assorted plesiosaur species (including diagrams, photographs, and videos), as well as extant marine tetrapods. These included observations/motion studies of extant marine tetrapods (and any anatomical constraints imposed by same).

Eliminating sea lion and rowing models

Our findings concurred at least in part with a previous study, which stated both the "sea lion" and "rowing" models [2, 19, 20] for plesiosaurs are contraindicated. As stated by reference 5: "Our analysis demonstrates that the sea lion- and rowing- models [for plesiosaurs] are kinematically impossible due to the prominent glenoid processes that would restrict the necessary posterior motion. Anatomical comparisons with extant tetrapod skeletons show that the glenohumeral joint of plesiosaurs is closest to that of underwater fliers such as sea turtles and penguins."

As can be observed in this still frame (dorsal view) from one of Ava's wireframe test runs (Figure 10),

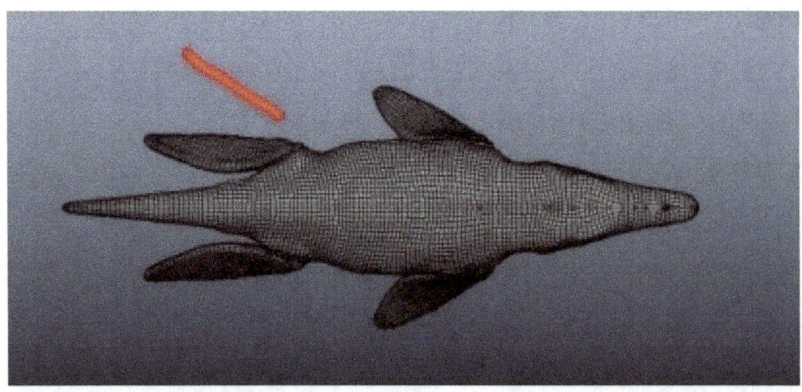

Figure 10. Virtual pliosaur Ava showing contraindicated ranges of motion (red bar) for posterior paddles.

the posterior
range of motion of the paddles, particularly the rear set, far exceeds feasible/safe ranges of motion for the animal and, if enacted, would have resulted in the paddle being partially, if not completely, dislocated at the femoral insertion point. A far more likely posterior range of motion for the rear paddles is indicated in red. This posterior range of motion is in line with previously published estimates for maximum anteroposterior ranges of motion for the rear paddles, i.e. approximately -57 degrees [21]. A more anatomically correct posterior range for the paddles (the posterior set in particular) can be seen in the wireframe still of Ava (Figure 11).

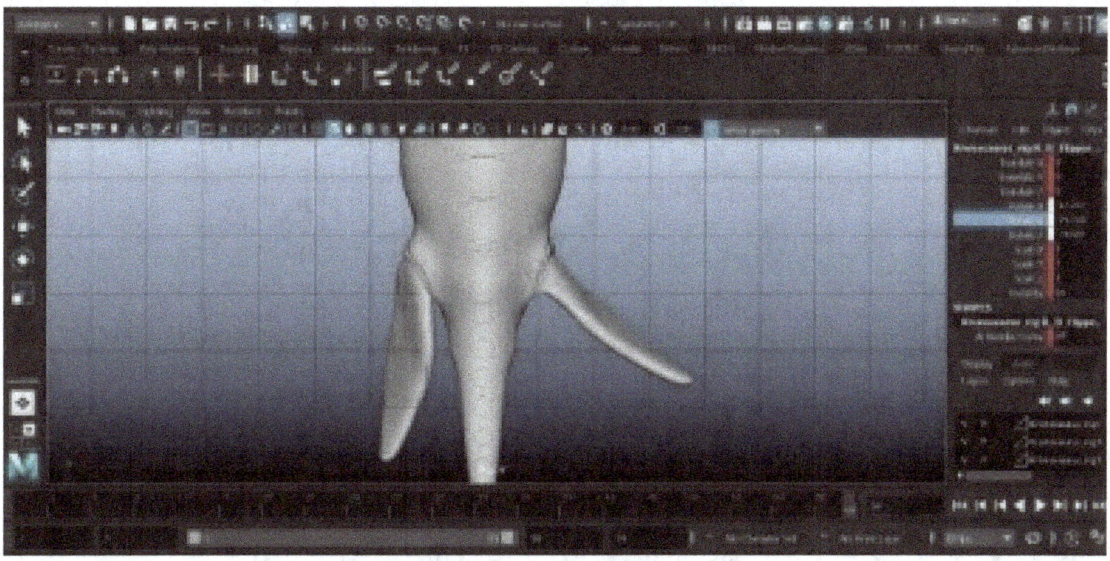

Figure 11. Dorsal view of Ava showing anatomically correct range of motion for right-side posterior paddle.

Increased posterior range of motion for the rear flippers

As previously stated, when observing both the caudal-oriented up-angling of the plesiosaur pelvis, as well as the adaptive shortening of the sacral ribs to accommodate the accompanying muscles, we calculated that the range of motion for the posterior paddles would be restricted in terms of completing downward thrusts. At the same time, however, we calculated that this "alcove" of adaptively-shortened ribs and up-angling of the pelvis would provide an elevated range of motion that would enable the animal to raise its rear flippers higher in the water than the anterior set. (Note: this increased posterior range of motion is supported, in part at least, by reference 5: "Both taxa show a large anteroposterior arc of motion for the hind limb, which may indicate that the hind limbs were important in maneuvering").

Our findings strongly suggest that this anteroposterior arc was not, in fact, primarily for maneuvering, but was, instead, mainly for locomotion. We calculate the resting (neutral) position of the rear paddles was approximately -24 degrees from horizontal (variation amongst different plesiosaur species applies) with dorsoventral ranges estimated at -0 to -37. These ranges align with previous findings [5, 21], particularly with the former study. When describing the pelvis of Thalassomedon haningtoni (DMNH 1588), Carpenter et al. [5] successfully demonstrated how the ilium overhung the acetabulum. "This overhang is sufficient to prevent the femur from rising above the horizontal. The same restriction is a common feature of plesiosaurs."

With Ava's posterior paddles able to travel higher in the water throughout their positive stroke (with the femurs at or near horizontal), they axially rotate during recovery, then push back hard through the water during the positive stroke, with the trailing edge of the paddles being near-vertical. The posterior paddles end in a far more lateral termination point than the anterior set. In living plesiosaurs this served the purpose of encompassing a different plane of motion than the front flippers, thereby pushing through different water than he front pair, and providing both improved acceleration and an overall increase in speed.

A differing range of motion for the anterior paddles

Unlike the posterior paddles, which moved largely laterally, the anterior paddles engaged in a range of motion more similar to that of extant penguins and sea turtles (but without the latter's ability to achieve extreme dorsal elevation due to the presence of an elbow joint). It is interesting to note that the underwater flight ranges of motion Ava exhibits in her anterior paddles was originally suggested by the Victorian-era paleontologists De La Beche and Conybeare [22], who stated that "the employment and motion of the paddle in swimming has many points of agreement with that of the wing in flight".

We calculate that the resting (neutral) position of the anterior paddles was approximately -30 degrees from horizontal (variation amongst different plesiosaur species again applies) with dorsoventral ranges estimated at +30 to -40. Our dorsal ranges for the fore-flippers align with previous findings [5] as do our ventral ranges [21]. During the recovery (negative) stroke Ava's fore-flippers exhibited axial rotation that enabled the leading edge to sweep forward and up, cutting the water efficiently as they rose. Unlike the rear paddles, when preparing for a power (positive) stroke the front pair were able to rise up above the midline. From there, they were able to push powerfully down and back, encompassing water separate from that affected by the rear flippers and, with both pairs working in conjunction, giving the animal increased speed and maneuverability.

Flippers working in conjunction (vacuum generation)

Our findings strongly suggest that, although both pairs of plesiosaur flippers moved simultaneously during the positive/power stroke, during the negative stroke the movements were most effective when they became slightly semi-synchronous. Studies of Ava's movements during her assorted versions supported our earlier hypotheses [23, 24] that the posterior flippers received some benefit from the anterior set. Although the paddles moved through separate planes of motion during the positive (power) stroke, they traveled though similar, close to horizontal ranges during the negative (recovery) stroke. Having the fore-flippers begin their return stroke a split-second before the hind flippers resulted in the leading edge of the fore-flippers pushing water and creating a veritable "vacuum" behind them. This turbulence (akin to the pressure wave dolphins ride ahead of moving ships) would have reduced the friction the posterior paddles experienced during their recovery stroke. In fact, during high-speed "sprints" (such as avoiding a predator or pursuing prey) the power generated by the fore-flippers would have generated enough suction that the posterior pair could virtually "come along for the ride". This allowed the rear paddles to easily catch up to the front paddles in time for a synchronous power (positive) stroke, let them conserve strength and absorb kinetic energy, and allowed for the generation of a more powerful positive stroke.

Inherent flexibility of plesiosaur paddles

Most previous plesiosaur locomotion studies have assumed that plesiosaurs had rigid or semi-rigid paddles [6]. In fact, the initial versions of plesiosaur swim cycle animations generated for this study [16] shared this assumption. However, our subsequent research, including a hands-on examination/comparison of several living extant sea turtle specimens and an analysis of the rubbery trailing edges of their paddles–tissue that extends far beyond the phalanges–is in accord with suggestions [5] that plesiosaur flippers were inherently flexible. Plesiosaur limbs tend to exhibit considerable hyperphalangy, even more so than extant sea turtles, with the phalanges exhibiting a regular reduction in size to the distal tip of the limb (Figure 4).

Figure 12. Distal portions of *Cryptoclidus* phalanges and radius, showing the attachment points of numerous ligaments (from the de la Salle Collection).

In addition, we noted during our examination of plesiosaur flipper fossils (Figure 12) that each phalange is joined to its neighbor by an assortment of ligaments. This plethora of phalanges, along with an abundance of connective tissue, led to both stability and flexibility, along with the ability to feather the limb during parts of the stroke (probably controlled by tendons). This is supported by the previous study [5] which stated: "In plesiosaurs, there is a decrease in the size of phalanges distally, which makes the flipper more flexible distally. This flexibility allows chord-wise flexing with a wave moving distally along the flipper. Many flying and swimming tetrapods can actively control bending and flexing of their wings or flippers with their limb muscles. This manipulation in cetaceans, penguins, sea lions, and sea turtles is mostly by tendons because the distal muscles are reduced or absent [24-27], and this reduction must have been true for plesiosaurs as well."

Although our earliest model had semi-rigid flippers, we found that Ava's flippers performed most effectively during positive (power) strokes when they were endowed with flexibility more in line with that of extant Chelonioidae (as opposed to the more rigid paddles of Spheniscidae). Ava's paddles were adjusted so that they demonstrated a change in shape when under load, becoming "feathered" (backward curving). It should be noted that this type of torsional flexibility is exhibited in sea turtles. During our model's strokes a flexion wave was added that traveled down the limb, producing a powerful flick at the end. Again, this aligns with and confirms the findings of the Carpenter et al. study [5], which refers to the flippers as having a torsional wave (akin to that of sea turtles) that travels down the length of each flipper during the positive stroke and creates additional thrust via a flick at the culmination of each stroke. This was further demonstrated by the use of human swimmers, swimming in tandem and equipped with slightly flexible plastic sheets to simulate the paddles of plesiosaurs. As aside note,

it bears mentioning that some of the Victorian researchers [22, 28] were potentially more accurate in their comparisons of plesiosaurs with extant animals than many "modern" paleontologists. In 1824 Conybeare noted [1] that in "its motion this animal must have resembled the [sea] turtle more than any other."

Effects of up-thrusts

It should be noted that sea turtles and penguins are able to change direction at will, both dorsally and ventrally, by simply altering the angle of thrust provided by their powerful fore-flippers. Sea turtles, in particular, with their more flexible paddles [14], are able to perform myriad and graceful maneuvers, suspending in place and shifting position in almost any direction they wish, and often by incorporating just one flipper at a time. With plesiosaurs having four well-developed flippers and an extensive system of phalanges/hyperphalanges, we can safely assume that they were capable of doing this as well.

Taking these factors into account, Ava was imbued with one additional effect for added realism; we added an approximation of how the pliosaur's body would respond when exposed to the strain of up-thrusts. When using the fore-flippers (or all four flippers, depending on positioning) to approach the water surface to spout or breathe (a regular occurrence taking place hundreds of times per day) we calculated that the marked increase in water resistance, caused by the sudden shift in stroke angle, would have had a direct effect on the animal's body. The skeletomuscular system would have had to compensate for the sudden and increased resistance during the start of its ascent. We see this skeletomuscular compensation exhibited by both penguins and sea turtles (more apparent in the former, due to the inherent obstruction caused by the latter's shell). Ava was designed to replicate the plesiosaur body's physiological response during its positive (power) strokes by the addition of a stiffening of the spine and a contraction of both the shoulder and abdominal muscles (anchored by robust gastralia). Neck flexion was incorporated to help keep the animal's head/eyes on target, with the head/neck helping to change direction by acting as a rudder.

A breakdown of the strokes

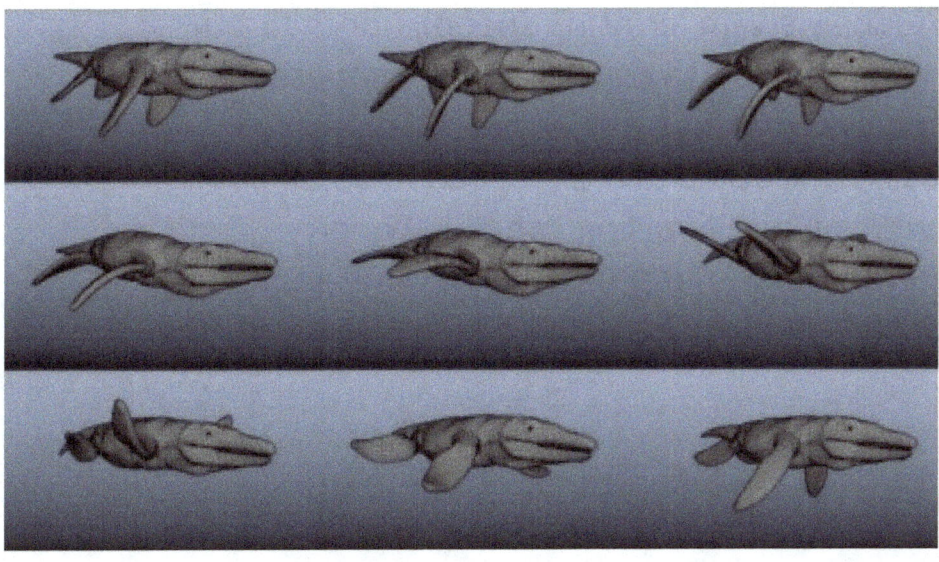

Figure 13. Wireframe swim cycle of Ava (animation by Mathieu Lafreniere).

In addition, a selection of frames from the animation is available for view (Figure 13). Starting point is top left and moving to the right row by row, typewriter-style (completion of the synchronous stroke takes place in the bottom row, far right).

A discussion on speed

Although calculations of velocity based on our research have been relegated to a future study, the following notations are provided for consideration. When contemplating the potential speed advantage of a single pair of paddles (as in sea turtles) versus a dual pair (as in plesiosaurs), the most basic analogy would be a comparison of the current world records in sculling–single sculls versus double sculls (currently Robert Manson at 6:30.74 vs Martin and Valent Sinković at 5:59.72). As these statistics demonstrate, the use of dual sets of paddles (rowers) results in a speed increase of only 12%. This is due to the fact that the sculler's oars are traveling through the same water/plane, hence they are subject to the Principal of Flipper Redundancy.

Granted, scullers are not plesiosaurs and, as Carpenter et al. noted [5]: "Despite some estimates of speed for plesiosaurs by Massare [29], we really do not know how fast plesiosaurs routinely swam (cruising speed), nor their burst speed." If we consider, say, the speed of the leatherback sea turtle (Dermochelys coriacea), this largest of extant chelonians has been clocked at a top speed of 35.28 km/h (21.92 mph), with a typical cruising speed between 1.80–10.08 km/h (1.12–6.26 mph) [30]. Given the active piscivorous lifestyle of most warm-blooded plesiosaurs, the speediness and agility of their prey (small fish and squid or, in the case of macrophagous pliosaurs, large fish, sharks, and even other marine reptiles) versus the inertness of the jellyfish that the endothermic leatherback preys upon, and taking into account their need to employ bursts of speed to either obtain prey or avoid becoming it, it is reasonable to assume that their upper-end speeds met or exceeded that of the leatherback. This is further supported by an analysis of the fusiform body-shape of the recently discovered polycotylid plesiosaur Mauriciosaurus, and a comparison between it and the leatherback turtle: "Nonetheless, the long and slim limbs that likely all participated in thrust generation, may have allowed for higher velocities" [31].

7. Conclusions

The advanced four-flippered swimming style of plesiosaurs was a great success. Plesiosaurs' unique mode of locomotion supported a variety of forms, from nimble pinniped-like polycotylids to outrageously long-necked elasmosaurs to macropredatory pliosaurs that could exceed extant baleen whales in mass.

Analysis of plesiosaur swim dynamics by means of a digital 3D armature (wireframe "skeleton") of a pliosauromorph ("Ava") demonstrates that: 1, plesiosaurs used all four flippers for primary propulsion; 2, plesiosaurs utilized all four flippers simultaneously; 3, respective pairs of flippers of Plesiosauridae, front and rear, traveled through distinctive, separate planes of motion, and; 4, the ability to utilize all four paddles simultaneously allowed these largely predatory marine reptiles to achieve a significant increase in acceleration and speed, which, in turn, contributed to their dominance in marine habitats for over 160 million years.

8. Appendix I: Specimens used in this study

The following real specimens or casts of plesiosaur skeletal remains were used in this study:

Cryptoclidus oxoniensis (AMNH 995), Elasmosaurus platyurus (ANSP 10081), Peloneustes philarchus, Cryptoclidus sp. (from the de la Salle Collection), Zarafasaura oceanis (from the Hawthorne Collection).

Photographs or renderings of the skeletons of the following genera of plesiosaur were also used: Myerasaurus, Cryptoclidus, Elasmosaurus, Rhomaleosaurus, Plesiosaurus, Archelon, Kronosaurus, Liopleurodon, Trinacromerum, Peloneustes, Thalassomedon, Maresaurus, and Mauriciosaurus. Photographs of the following modern skeletons were used: Zalophus californianus, Dermochelys coriacea, Chelonia mydas, and Spheniscus magellanicus, as was video footage of the following extant animals: Zalophus californianus, Dermochelys coriacea, Chelonia mydas, Pygoscelis adeliae, and Eretmochelys imbricata. Lastly, physical examinations of the paddles and limb girdles of the following live specimens of sea turtles and penguins were performed as part of this study: Lepidochelys kempii, Chelonia mydas, and Aptenodytes patagonicus.

Acknowledgements

We wish to express our gratitude to computer animator Mathieu Lafreniere, for his unparalleled skill at bringing our virtual pliosaur to life. We thank him not only for his insight and ability, but also his patience and tenacity as we worked hand in hand to recreate a style of underwater flight that hasn't been seen in over 65 million years. Thanks also to Adam Smith for access to the plesiosaur images, Tim Delaney of SeaWorld for a behind-the-scenes tour, Jose Patricio O'Gorman for use of his plesiosaur rib illustration, and Ryosuke Motani for the use of illustrations featured in Plesiosaurs Neo.

Figures

REFERENCES

[1] Conybeare, W.D. On the discovery of an almost perfect skeleton of the plesiosaurus. Transactions of the Geological Society of London 1824, S2, 381-389.

[2] Watson, D.M.S. The elasmosaurid shoulder-girdle and fore-limb. Proceedings of the Zoological Society of London 1924, 28, 85-95,.

[3] Robinson, J.A. The locomotion of plesiosaurs. Neues Jahrbuch für Paläontologie, Abhandlugen 1975 149, 286-332,

[4] Robinson, J.A. 1977. Intracorporal force transmission in plesiosaurs. Neues Jahrbuch für Paläontologie, Abhandlugen 1977, 153, 86-128,

[5] Carpenter, K.; Sanders, F.; Reed, B.; Reed, J.; Larson, P. Plesiosaur swimming as interpreted from skeletal analysis and experimental results. Transactions of the Kansas Academy of Science 2010, 113, 1-34.

[6] Muscutt, L.E.; Dyke, G.; Weymouth, G.D.; Naish, D.; Palmer, C.; Ganapathisubramani, B. The four-flipper swimming method of plesiosaurs enabled efficient and effective locomotion. Proceedings of the Royal Society B, Biological Sciences 2017, 284, doi: 10.1098/rspb.2017.0951.

[7] McMenamin, M. Permian Aquatic Reptiles. PaleorXiv 2019, doi:10.31233/osf.io/wb6h7.

[8] Caldwell, M.W. Modified perichondral ossification and the evolution of paddle-like limbs in ichthyosaurs and plesiosaurs. Journal of Vertebrate Paleontology 1997, 17, 534-547.

[9] Taylor, M.A. The lifestyle of plesiosaurs. Nature 1986, 319, 179.

[10] Caldwell, M.W. From fins to limbs to fins: limb evolution in fossil marine reptiles. American Journal of Medical Genetics 2002, 112, 236-249.

[11] Massare, J.A. Swimming capabilities of Mesozoic marine reptiles: a review. In Mechanics and Physiology of Animal Swimming; Maddock, L.; Bone, Q.; Rayner, J.M.V., Eds.; Cambridge University Press, Cambridge, UK, 1994, pp. 133-149.

[12] Rothschild, B.M.; Clark, N.D.; Clark, C.M. Evidence for survival in a Middle Jurassic plesiosaur with a humeral pathology: what can we infer of plesiosaur behavior? Palaeontologia Electronica 2018, 21, 1-11.

[13] Flammang, B.E.; Suvarnaraksha, A.; Markiewicz, J.; Soares, D. Tetrapod-like pelvic girdle in a walking cavefish. Scientific Reports 2016, 6, 23711.

[14] Song, S.H.; Kim, M.S.; Rodrigue, H.; Lee, J.Y.; Shim, J.E.; Kim, M.C.; Chu, W.S.; Ahn, S.H. Turtle mimetic soft robot with two swimming gaits. Bioinspiration and Biomimetics 2016, 11, 036010

[15] Long, J.H.; Schumacher, J.; Livingston, N.; Kemp, M. Four flippers or two? Tetrapodal swimming with an aquatic robot. Bioinspiration & Biomimetics 2006 1, 20-29.

[16] Buchy, M.C.; Metayer, F.; Frey, E.; Osteology of Manemergus anguirostris n. gen. et sp., a new plesiosaur (Reptilia, Sauropterygia) from the Upper Cretaceous of Morocco. Palaeontographica Abteilung A Palaozoologie Stratigraphie 2005 272, 97-120.

[17] Taylor, G.K.; Nudds, R.L.; Thomas, A.L. Flying and swimming animals cruise at a Strouhal number tuned for high power efficiency. Nature 2003, 425, 707-711.

[18] Sanders, F.; Carpenter, K.; Reed, B.; Reed, J. Plesiosaur swimming reconstructed from skeletal analysis and experimental results. Transactions of the Kansas Academy of Science 2010, 113, 1–34.

[19] Tarlo, L.B. The scapula of Pliosaurus macromerus Phillips. Palaeontology 1958, 1, 193-199.

[20] Godfrey, S.J.. Plesiosaur subaqueous locomotion. Neues Jahrbuch für Geologie und Paläontologie, Monatshefte 1984, 11, 661-672.

[21] Liu, S.; Smith, A.S.; Gu, Y.; Tan, J.; Karen Liu, C.; Turk, G. Computer simulations imply forelimb-dominated underwater flight in plesiosaurs. PLOS Computational Biology 2015, https://doi.org/10.1371/journal.pcbi.1004605.

[22] De la Beche, H.; Conybeare, W.D. Notice of the discovery of a new fossil animal, forming a link between the Ichthyosaurus and crocodile, together with general remarks on the osteology of the Ichthyosaurus. Transactions of the Geological Society of London 1821, 5, 559-594.

[23] Hawthorne, M. Plesiosaur facts: Creatures proper swimming method, 2015, https://www.kronosrising.com/proper-plesiosaur-swimmingmethods-max-hawthorne/

[24] Hawthorne, M.; McMenamin, M. Plesiosaur swimming method—A flipper propulsion study

https://www.kronosrising.com/plesiosaurs-swam-flipper-propulsion-study/

[25] Clark, B.D; Bemis, W. Kinematics of swimming of penguins at the Detroit Zoo. Journal of Zoology, London 1979, 188, 411-428.

[26] Schreiweis, D.O. A comparative study of the appendicular musculature of penguins (Aves: Sphenisciformes). Smithsonian Contributions to Zoology 1982, 341, 1-46.

[27] Fish, F. Structure and mechanics of nonpiscine control surfaces. IEEE Journal of Oceanic Engineering 2004, 29, 605-621.

[28] Seeley, H.G.. Note on some of the generic modifications of the plesiosaurian pectoral arch. Quarterly Journal of the Geological Society 1874, 120, 436-449.

[29] Massare, J.A.. Swimming capabilities of Mesozoic marine reptiles: Implications for method of predation. Paleobiology 1988, 14, 187-205.

[30] Shweky, Rachel, 1999, The Physics Factbook, https://hypertextbook.com/facts/1999/RachelShweky.shtml.

[31] Frey, E.; Mulder, E.W.A.; Stinnesbeck, W.; Rivera-sylva, H.E.; Padilla-Gutiérrez, J.M.; González-González, A.H. A new polycotylid plesiosaur with extensive soft tissue preservation from the early Late Cretaceous of northeast Mexico. Boletín de la Sociedad Geológica Mexicana 2017, 69, http://www.scielo.org.mx/scielo.php?script=sci_arttext&pid=S1405-33222017000100087#aff3

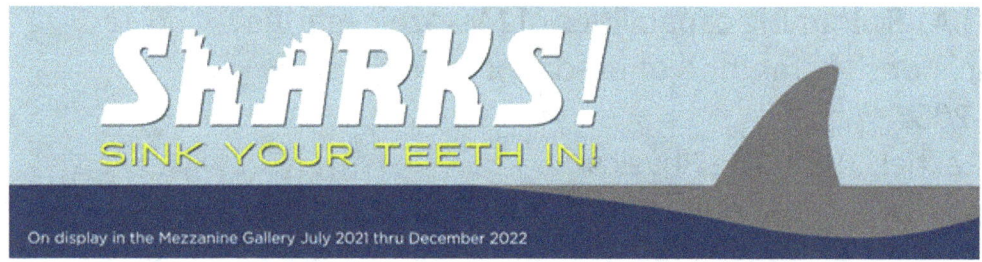

On display in the Mezzanine Gallery July 2021 thru December 2022

The Dinosaur Farm

https://www.dinosaurfarm.com/

Toys * Books* and all things dinosaur

Never heard of the Jurassic World Dominion Atrociraptor? Don't fret, we've got you covered!

By Dr. Brian Curtice
What's an...*Atrociraptor*?

Making its big-screen debut in Jurassic World - Dominion is *Atrociraptor!* The name means "savage robber". It was discovered in 1995 and named in 2004 based on a partial skull found in 68 million-year-old Late Cretaceous rocks of Canada near Drumheller, Alberta. The bones have features

Atrociraptor marshalli Currie & Varricchio 2004

Atrociraptor

found only in the "raptors", the group of dromaeosaurid dinosaurs closely related to birds.

For those curious, the animal is known from a right maxilla (upper jaw) and a pair of premaxillae (snout) and dentaries (lower jaws), plus some associated teeth and "bone fragments" as Currie and Varricchio describe the remaining bones. The fact it took 10 years from discovery to being named is not unusual in paleontology as researchers are often working on other projects when a new specimen appears, the fossil then has to be prepared for study, and the study itself takes quite a while.

Atrociraptor lived at the same time as *Bambiraptor*, *Saurornitholestes*, and *Velociraptor* but differs from them in having a face that is short and deep and teeth that are all roughly the same size and point throatward.

At 6' long nose to tail, and around 30 lbs (presuming the body matches the skull of course!) this likely feathered predator was the size of a turkey but with teeth and claws that make similarly-weighted felids such as ocelots, caracals, or servals jealous!

Atrociraptor

Kingdom: Animalia
Phylum: Chordata
Clade: Dinosauria
Clade: Saurischia
Clade: Theropoda
Family: †Dromaeosauridae
Clade: †Eudromaeosauria
Subfamily: †Saurornitholestinae
Genus: †Atrociraptor
Species: †A. marshalli
Binomial name †Atrociraptor marshalli

Currie & Varricchio, 2004

FossilCrates.com

The Strange and unusual Diplocaulus

**By Shetan Noir
with credit to Wikipedia**
https://en.m.wikipedia.org/wiki/Diplocaulus

In the prehistoric world there are many unusual looking creatures, but none more strange or unique than the amphibian Diplocaulus. It's odd looking boomerang shaped head gives it it's name
Diplocaulus, which means "double caul".

It's suggested that it may have looked like a salamander body wise but that head sets it apart.

Diplocaulus was a Piscivore that fed on fish and insects that ventured to close to it in the water.

The Diplocaulus lived and thrived in watery swamp areas and avoided predators like DIMETRODON by burrowing into lairs under the mud.

But despite its best efforts some Diplocaulus did meet thier end from bites by dimetrodons as several Diplocaulus skull fossils have been found with chunks missing from the skulls.

The Diplocaulus skull might have aided it in the water as some scientists believe that its head may have been used as a hydrofoil. This would have made it possible for it to skim on top of a body of water. If that's true, then the shape of its head could have also been used as a way for this animal to steer itself. This would have allowed it to traverse streams, rivers and lakes with ease.

The Diplocaulus is an extinct genus of lepospondyl amphibians which lived from the Late Carboniferous to the Late Permian in North America and Africa.

According to the Wikipedia page on Diplocaulus,

The *Diplocaulus* is by far the largest and best-known of the lepospondyls, characterized by a distinctive boomerang-shaped skull.

Remains attributed to *Diplocaulus* have been found from the Late Permian of Morocco and represent the youngest-known occurrence of a lepospondyl.

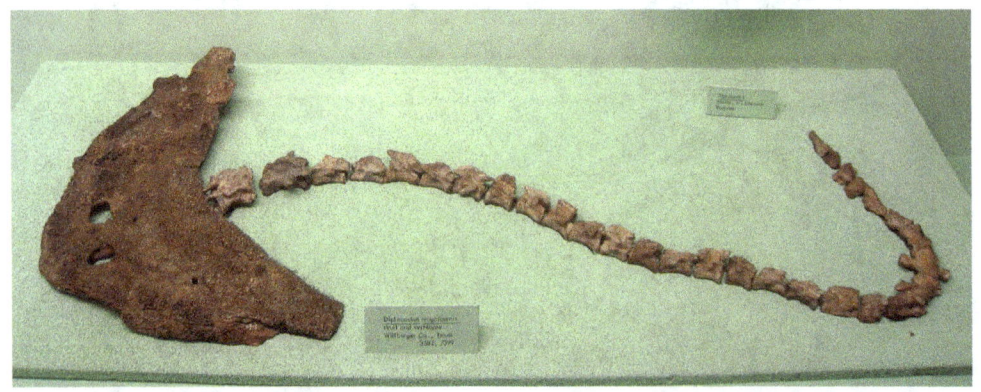

Diplocaulus had a stocky, salamander-like body, but was relatively large, reaching up to 1 m/3.3 ft in length and weighing up to 10lbs. This would be comparable to today's Chinese giant salamanders and Japanese giant salamanders who range in size of 3 to 5ft in length.

Although a complete tail is unknown for the genus, a nearly complete articulated skeleton described in 1917 preserved a row of tail vertebrae near the head. This was construed as circumstantial evidence for a long, thin tail capable of reaching the head when and if the animal was in a curled up position. Most studies since this discovery have argued that anguiliform (eel-like) tail movement was the main force of locomotion utilized by *Diplocaulus* and its relatives.

D. salamandroides

Meeting the relatives,
D. salamandroides was the first species of *Diplocaulus* to be discovered. Remains from this species were discovered near Danville, Illinois by William Gurley and J.C. Winslow, a pair of local geologists. The fossils were later described by renowned paleontologist Edward Drinker Cope in 1877. This species is only known from a small number of vertebrae sent to Cope by Gurley and Winslow. These vertebrae were noted for their similarities to those of salamanders (hence the specific name *salamandroides*), although Cope was reluctant to refer them to any known group. A large jaw bone with labyrinthodont teeth was associated with some of these vertebrae, but it was much larger than expected for the vertebrae and likely belonged to *Eryops* or some other larger amphibian.

D. salamandroides
could be distinguished from *D. magnicornis* by its small size (from a fifth to a sixth the size of the latter) and less pronounced accessory articular processes (at the time identified as zygosphene-zygantrum articulations).

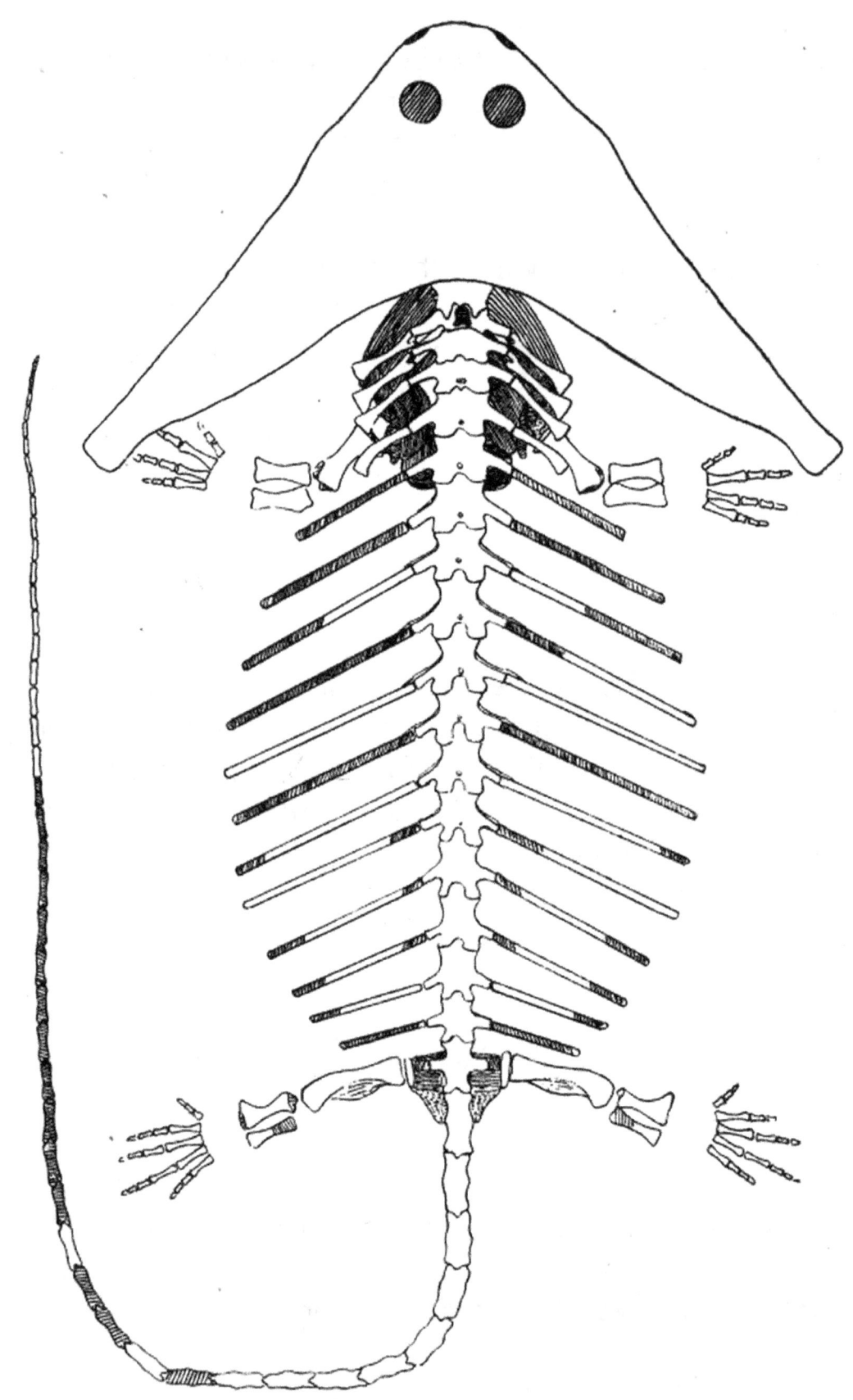

Fig. 4.—*Diplocaulus*, skeleton, restored. Barred parts restored. Pectoral girdle should be somewhat broader.

The rocks in which these fossils were discovered had been informally referred to as the "*Clepsydrops* shales", named after a local genus of early synapsid by Cope in 1865. The shales were initially believed to be from either the Permian or Triassic periods in age based on the purported presence of reptile and lungfish fossils. By 1878, Cope had decided that the site was Permian. In 1908, E.C. Case noted that the shales also contain remains from fish which were from the late Carboniferous and early Permian periods. He argued that, while the *Clepsydrops* shales of Illinois and the similar red beds of Texas were evidently formed after the major Carboniferous coal deposits, there was not sufficient evidence to exclude them from the Carboniferous period itself. Nowadays the *Clepsydrops* shales are typically assigned to the McLeansboro or Mattoon Formations. *D. salamandroides* fossils have also been found in Pennsylvania. These formations are now believed to be Missourian (late Carboniferous) in age.

D. magnicornis

This species, described by Cope in 1882, is by far the most common and well-described member of the genus. *D. magnicornis* was the first species known from more than vertebrae, and it allowed Cope and other paleontologists to realize the nature of *Diplocaulus* as a bizarre long-horned "batrachian" (amphibian). Much of modern knowledge on the genus is based on this species, as it outnumbers any other *Diplocaulus* remains by hundreds of specimens. *D. magnicornis* had a wide temporal distribution throughout the red beds of Texas and Oklahoma.

D. brevirostris

D. brevirostris was similar to *D. magnicornis*, although it was significantly more rare. It is represented by a small number of specimens found in an early strata of the Texas red beds, specifically the Arroyo Formation of the Clear Fork Group. This species can be differentiated from *D. magnicornis* by the much shorter and blunter snout compared to the length of the skull as a whole. In addition, the horns are more elongated, the parietals have a convex upper surface, and the rear edge of the skull is more strongly and smoothly curved. While juvenile members of *D. magnicornis* also have a smoothly curved rear edge of the skull, all known *D. brevirostris* specimens are clearly adults as shown by their robust skull ornamentation, long horns, and large size. Therefore, this trait is a legitimate distinguishing feature of adult specimens of this species. The only specimen known from more than a skull is the type specimen, AM 4470, which preserves some vertebrae similar to those of "*D. primigenius*". E.C. Olson, the original describer of the species, suggested that it occupied different habitats than *D. magnicornis* such as mountain streams, accounting for its comparative rarity. However, other studies have suggested that *D. magnicornis* would have lived in similar environments, invalidating Olson's hypothesis.

D. recurvatus

This species, from the Vale Formation of the Texas red beds, was very similar to *D. magnicornis*, and partially coexisted alongside that species in younger strata. Olson hypothesized that *D. recurvatus* may have been descended from an early stock of *D. magnicornis*.

D. recurvatus differs from *D. magnicornis* in one specific trait: the tips of the tabular horns are "crooked". The tips are bent relative to the rest of the horns, and abruptly taper. Comparison to a growth series of *D. magnicornis* indicates that *D. recurvatus* specimens had developmental pathways which significantly differed from *D. magnicornis*. For example, skull length and width seem to be inversely correlated in *D. recurvatus* and directly correlated in *D. magnicornis*. In addition, the restriction in the horns of *D. recurvatus* develops in an area which would otherwise expand in adult *D. magnicornis*.

Most of the big amphibians went extinct, including the boomerang-headed *Diplocaulus*.

D. minimus

Diplocaulus minimus is a species known from the Ikakern Formation of Morocco. It had an unusually asymmetrical skull, with the left prong being long and tapering as in other species but the right prong being much shorter and more rounded. This feature was present in multiple skulls referred to this species, so it is very unlikely to be a result of crushing or distortion. Some studies have suggested that this species is more closely related to *Diploceraspis* than to *Diplocaulus magnicornus*. This may suggest that either *Diplocaulus* is not a true monophyletic genus, that *Diploceraspis* is a junior synonym of the genus, or that *"Diplocaulus" minimus* represents a distinct genus.

Dubious species

D. limbatus was the third species of *Diplocaulus* to be named, and remained the second most well known member of the genus until the 1950s. It was described by E.D. Cope in 1895 based on several incomplete specimens found in the Texas red beds. The type specimen was a poorly preserved skull and partial skeleton designated AM 4471. Cope found that the skull of this specimen had shorter, thinner horns than those of *D. magnicornis*, as well as a seemingly unique feature: a large notch separating the quadratojugal from the rest of the tabular horn. E.C. Case later provided additional distinctions present in a skull referred to *D. limbatus*, including smoother edges to the skull, larger eyes, and more pointed horns. However, additional *D. limbatus* specimens prepared by Douthitt have shown that many of Case's identifications were erroneous, and that only the notch identified by Cope could be used to distinguish it from *D. magnicornis*. In 1951, E.C. Olson concluded that AM 4471 was too poorly preserved to differentiate from *D. magnicornis*, and therefore he designated *D. limbatus* as a synonym of that species. However, he also analyzed the referred *D. limbatus* skull described by Case, AM 4470, and found that it was unique enough to qualify as the type specimen of a new *Diplocaulus* species: *D. brevirostris*.

- *D. copei* and *D. pusillus* were both named by German paleontologist Ferdinand Broili in 1904. *D. copei* was known from three Texan specimens, all of which were heavily crushed and incomplete. Broili argued that this species was unique due to its small size and horns which bend inwards. However, E.C. Case could find no way to distinguish between its specimens and those of *D. magnicornis* and "*D. limbatus*", and he rejected the species as indeterminate, a decision followed by later sources. *D. pusillus,* known from a pair of minuscule skulls found in Texas and stored at the Palaeontological Museum of Munich, is a more controversial species. The skulls

- came from some other amphibian from the area, such as *Trimerorhachis*. In 1918, S.W. Williston used the *D. pusillus* specimens as the basis for *Platyops parvus*, a new genus of diplocaulid. In 1946, E.C. Case revised Williston's name to *Permoplatyops parvus*, as the genus name "*Platyops*" was already in use. He brought up the possibility that the skulls were from an extremely young *Diplocaulus*, and in response Olson (1951) designated *D. pusillus* (and therefore *Permoplatyops parvus*) as a synonym of one of the other red bed *Diplocaulus* species, such as *D. magnicornis*.

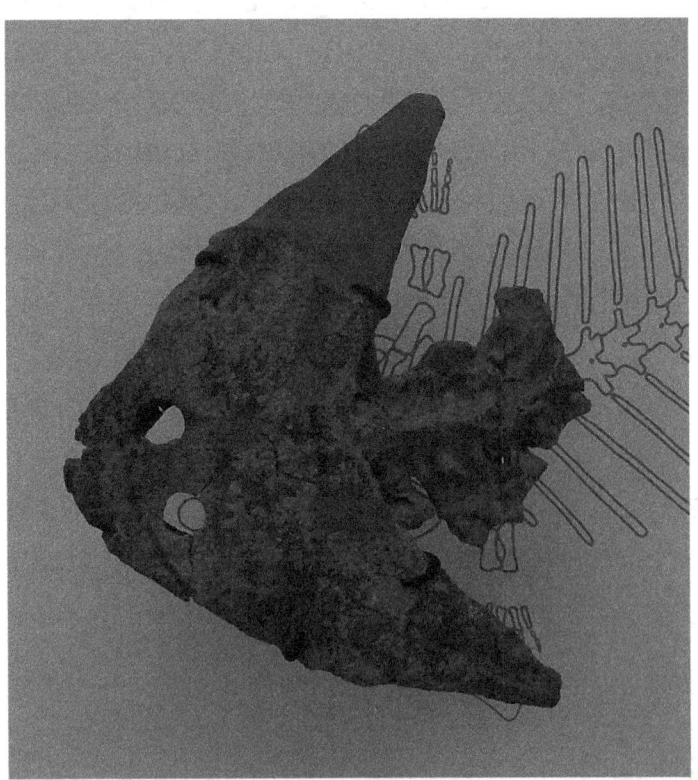

- *D. primigenius* was described in 1921 by M.G. Mehl based on a single specimen preserving a skull, shoulder elements, and a string of vertebrae. The skull was seemingly identical to that of *D. magnicornis*, but the vertebrae were peculiar. They were quite enlarged, particularly the neural spines which were tall, rough structures with a depression at their highest extent.[19] E.C. Olson (1951) noted that the vertebrae were comparable to those of the holotype of *D. brevirostris* (AM 4470), but also that the skull was much more akin to *D. magnicornis* instead. While Olson did decide to synonymize *D. primigenius* with *D. magnicornis*, he also noted that the specimen remained an interesting conundrum with implications for the disconnect between vertebral and skull development in *Diplocaulus*.

- *D. parvus*, named by E.C. Olson in 1972, was designated as a new species with no connection whatsoever with "*Permoplatyops parvus*", which at that point was treated as a synonym of *D. magnicornis*. *D. parvus* is known from a single specimen from the Chickasha Formation of Oklahoma. It was generally very similar to *D. recurvatus*, differing primarily due to its smaller size as isolated geographical location. Germain (2010) did not consider these traits sufficient enough to justify retaining *D. parvus* separate from *D. recurvatus*. The *D. parvus* specimen is potentially the youngest *Diplocaulus* fossil recovered from North America, at about 270 million years old.

- Douthitt, Herman (September 1917). "The Structure and Relationships of Diplocaulus" (PDF). *Contributions from Walker Museum*. **2**(1): 1–42.

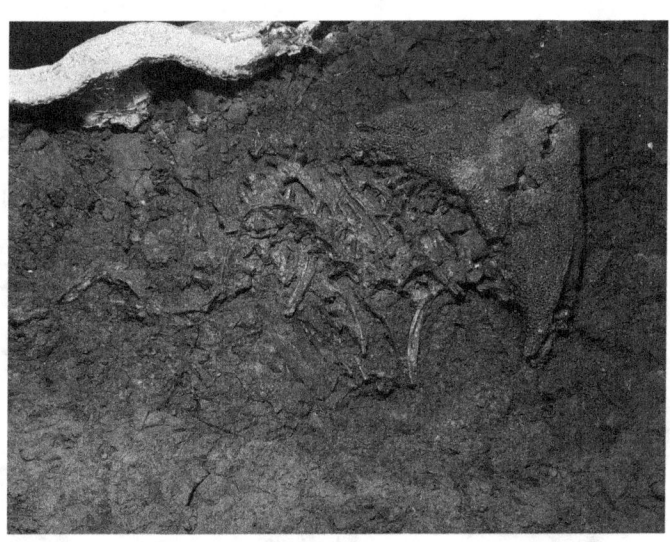

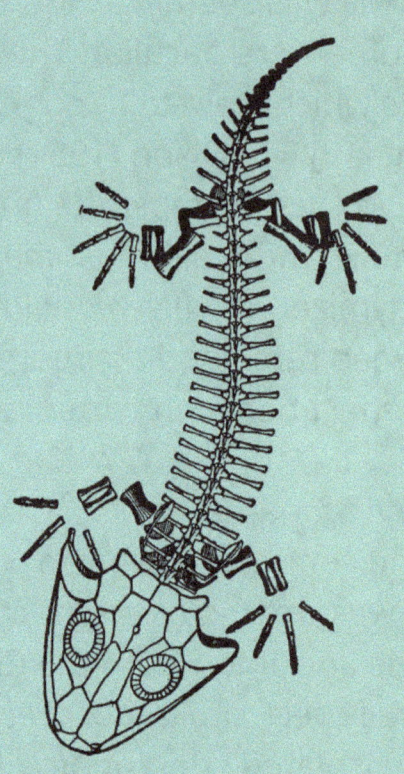

Diplocaulus

Kingdom: Animalia
Phylum: Chordata
Subclass: † Lepospondyli
Order: † Nectridea
Family: † Diplocaulidae
Genus: † Diplocaulus

- ^**abc**Beerbower, J.R. (November 1963)."Morphology, paleoecology, and phylogeny of the Permo-Pennsylvania amphibianDiploceraspis".Bulletin of the Museum of Comparative Zoology.**130**(2): 31–108.

- ^**abcdef**Cruickshank, A. R. I.; Skews, B. W. (1980). "The Functional Significance of Nectridean Tabular Horns (Amphibia: Lepospondyli)".Proceedings of the Royal Society B: Biological Sciences.**209**(1177): 513–537.doi:10.1098/rspb.1980.0110.S2CID110443064.

- ^**abcde**Case, E.C. (1911)."Revision of the Amphibia and Pisces of the Permian of North America".Carnegie Institution of Washington Publication.**146**: 15–91.

- ^**abcdefg**Olson, E.C. (12 January 1951)."Diplocaulus: A study in growth and variation".Fieldiana: Geology.**11**(2): 59–149.

- ^**abc**Olson, Everett C. (September 1972). "Diplocaulus parvusn. sp. (Amphibia: Nectridea) from the Chickasha Formation (Permian: Guadalupian) of Oklahoma".Journal of Paleontology.**46**(5): 656–659.

- ^**ab**Harris, Susan K.; Lucas, Spencer G.; Berman, David S.; Henrici, Amy C. (2005)."Diplocaulus cranial material from the lower Abo Formation (Wolfcampian) of New Mexico and the stratigraphic distribution of the genus".New Mexico Museum of Natural History and Science Bulletin.**30**: 101–103.

- ^Cope, E.D. (2 November 1877)."Descriptions of Extinct Vertebrata from the Permian and Triassic Formations of the United States".Proceedings of the American Philosophical Society.**17**(1): 182–193.

- ^Case, E.C. (1900)."Contributions from Walker Museum. I: The Vertebrates from the Permian Bone Bed of Vermilion County, Illinois".The Journal of Geology.**8**(8): 698–729.doi:10.1086/620866.

- Cope, E.D. (15 September 1882)."Third Contribution to the History of the Vertebrata of the Permian Formation of Texas".Proceedings of the American Philosophical Society.**20**(112): 447–461.

- ^**ab**Olson, Everret C. (November 1953). "Integrating Factors in Amphibian Skulls".The Journal of Geology.**61**(6): 557–568.doi:10.1086/626128.S2CID128813415.

- ^Olson, E.C. (27 June 1952)."Fauna of the upper Vale and Choza: 6,Diplocaulus".Fieldiana: Geology.**10**(14): 147–166.

- ^**ab**Germain, Damien (27 May 2010). "The Moroccan diplocaulid: the last lepospondyl, the single one on Gondwana".Historical Biology.**22**(1–3): 4–39.doi:10.1080/08912961003779678.S2CID128605530.

- ^Cope, E.D. (15 November 1895)."Some New Batrachia from the Permian of Texas".Proceedings of the American Philosophical Society.**34**: 452–457.

- ^Broili, Ferdinand (14 June 1904)."Permische Stegocephalen un Reptilien aus Texas".Palaeontographica.**51**: 1–120.

- ^**ab**Williston, S.W. (1909)."The Skull and Extremities ofDiplocaulus".Transactions of the Kansas Academy of Science.**22**: 122–132.doi:10.2307/3624731.JSTOR3624731.

- ^Case, E.C. (September 1946)."A Census of the Determinable Genera of the Stegocephalia".Transactions of the American Philosophical Society.**35**(4): 323–420.doi:10.2307/1005567.hdl:2027/mdp.39015071637537.JSTOR1005567.

- ^Mehl, M.G. (1921)."A New form of Diplocaulus".Journal of Geology.**29**(1): 48–56.doi:10.1086/622753.

- ^Palmer, D., ed. (1999).The Marshall Illustrated Encyclopedia of Dinosaurs and Prehistoric Animals. London: Marshall Editions. p. 55.ISBN978-1-84028-152-1.

- ^Zoehfeld, Weidner K.; Bakker, Robert T.; Flis, Chris J.; Pettersson, Carl B.; Bell, Troy H. (2013)."Abstract: BURROWS AND BREAK-INS ON THE TEXAS PERMIAN DELTA: STACKED AESTIVATING AMPHIBIANS AND ATTACKS BY DIMETRODON (2013 GSA Annual Meeting in Denver: 125th Anniversary of GSA (27-30 October 2013))".gsa.confex.com.

- ^Cf.

Exhibit for sale

Author Spotlight on Evan Johnson Ransom

OVER 1,200 AMAZING DINOSAURS, FAMOUS FOSSILS, AND THE LATEST DISCOVERIES FROM THE PREHISTORIC ERA

DINOSAUR WORLD

ILLUSTRATED BY JULIUS CSOTONYI • WRITTEN BY EVAN JOHNSON-RANSOM

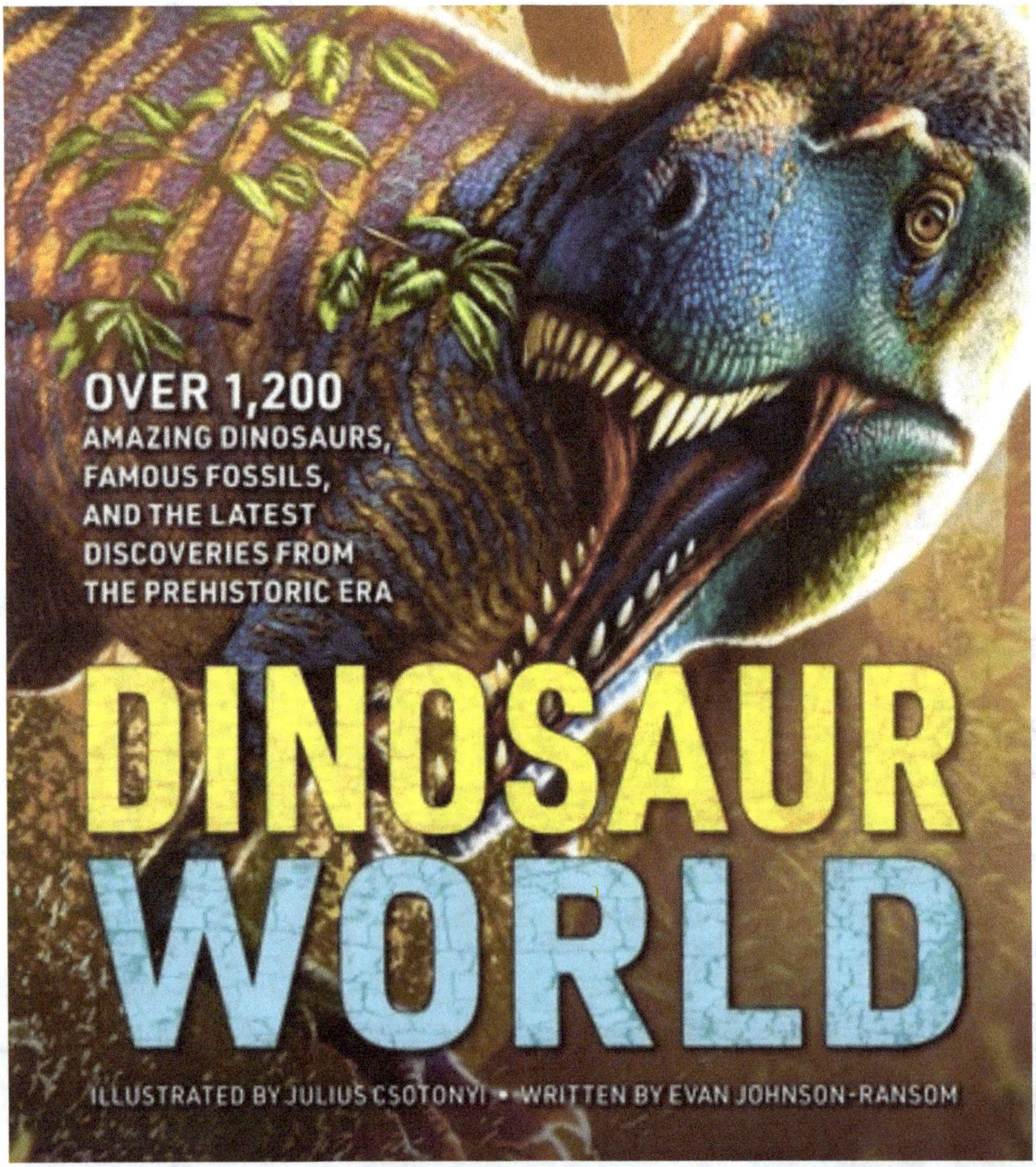

Author Spotlight
on
Evan Johnson Ransom

Simon & Schuster

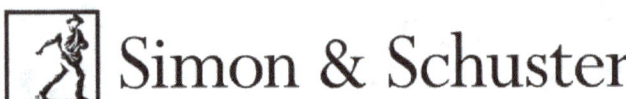

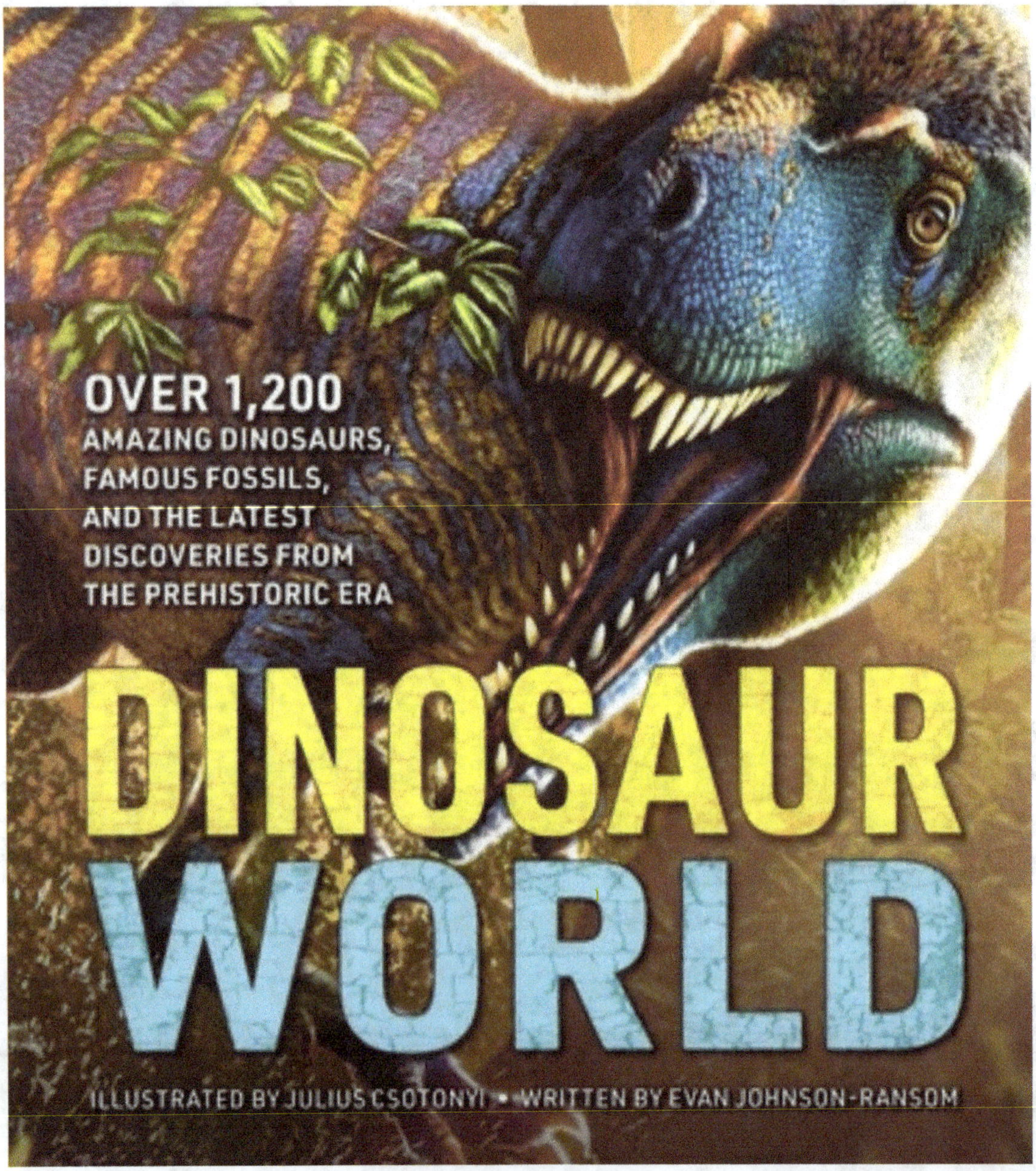
OVER 1,200
AMAZING DINOSAURS, FAMOUS FOSSILS, AND THE LATEST DISCOVERIES FROM THE PREHISTORIC ERA
DINOSAUR
WORLD
ILLUSTRATED BY JULIUS CSOTONYI • WRITTEN BY EVAN JOHNSON-RANSOM

Can you please tell us more about your background?

My name is Evan and I am a vertebrate paleontologist. I am currently enrolled at the University of Chicago for my PhD. I received my Master's at Oklahoma State University Center for Health Sciences in 2021, and my Bachelor's from DePaul University in 2021.

How did you become interested in Dinosaurs?

I've been interested in dinosaurs since I was 2 years old. When I was in preschool I always cried when my mother dropped me off in the morning. She noticed my classmates would console me with dinosaur toys. That year for my birthday and Christmas, my family gave me dinosaur toys and I was quickly enamored by dinosaurs. My grandmother and mother took me to dinosaur museums, in addition to me watching dinosaur documentaries, and reading books about dinosaurs.

Can you tell us more about your book, Dinosaur World?

The book details information on every dinosaur that has been discovered to date with information on their size, diet, ecology, and significance in paleontology. In addition to the dinosaur descriptions, the book also talks about the biology of dinosaurs (feeding behavior and anatomy), museums that house dinosaur specimens and biographies of prolific and aspiring paleontologists from diverse and under-represented backgrounds.

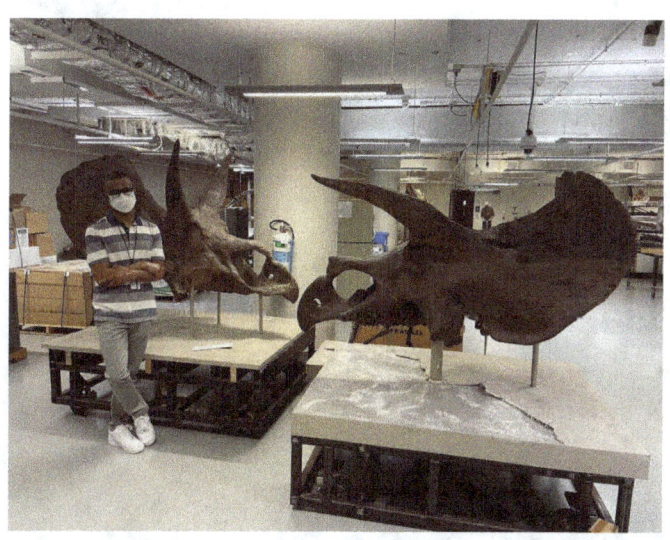

Is this the first book you have written?

Dinosaur World is the first book that I've written.

How did you start the research for your book?

I mostly looked at my old dinosaur books from childhood for references. One particular book that has been a part of my childhood was "The Complete Book of Dinosaurs" by Douglas Dixon, which was published in 2006. The reason I refer to it as my favorite dinosaur book was that it had illustrations and information of every dinosaur that was discovered at that time. I've always enjoyed reading books that provide detailed information on a dinosaurs

Where can your current book be purchased from?

The book can be pre-ordered on websites such as Amazon, Barnes and Noble, Books a Million, Simon & Schuster, and Indie Bound. It does have an alternative title called "The Greatest Dinosaur Book Ever", which was the original name before being changed to "Dinosaur World". It's best to search for the book by its ISBN: 9781646433162. You can preorder the book now and it will be released February 2023.

Will you have more books coming out?

I don't have any more books coming out at the moment. I just started my PHD program and am currently focusing on my research.

What has been your favorite dinosaurs to research?

My favorite dinosaurs to research are the theropod dinosaurs like Tyrannosaurus rex, Allosaurus, Spinosaurus, and Velociraptor. I've always been interested in the feeding behavior of predatory dinosaurs and the roles they play in their environment.

Do you have any social media sites or a website?

I have a Twitter account, @EJR_Paleo_MSc.